The mechanisms of macroevolutionary change have long been a contentious issue. Palaeoecological evidence, presented in this book, shows that evolutionary processes visible in ecological time do not build up into macroevolutionary trends, contrary to Darwin's original thesis.

The author discusses how climatic oscillations on ice-age time-scales are paced by variations in the Earth's orbit, and have thus been a permanent feature of Earth history. There is, however, little evidence for macroevolutionary change in response to these climatic changes, suggesting that over geological time macroevolution does not occur as a result of accumulated short-term processes. These conclusions are used to construct a post-modern evolutionary synthesis in which evolution and ecology play an equal role.

Written by a leading palaeoecologist, this book will be of interest to researchers in both ecology and evolutionary biology.

Evolution and Ecology: The Pace of Life

Cambridge Studies in Ecology presents balanced, comprehensive, up-to-date, and critical reviews of selected topics within ecology, both botanical and zoological. The Series is aimed at advanced final-year undergraduates, graduate students, researchers and university teachers, as well as ecologists in industry and government research.

It encompasses a wide range of approaches and spatial, temporal, and taxonomic scales in ecology, experimental, behavioural and evolutionary studies. The emphasis throughout is on ecology related to the real world of plants and animals in the field rather than on purely theoretical abstractions and mathematical models. Some books in the Series attempt to challenge existing ecological paradigms and present new concepts, empirical or theoretical models, and testable hypotheses. Others attempt to explore new approaches and present syntheses on topics of considerable importance ecologically which cut across the conventional but artifical boundaries within the science of ecology.

CAMBRIDGE STUDIES IN ECOLOGY

Series Editors
H. J. B. Birks   *Botanical Institute, University of Bergen, Norway, and Environmental Change Research Centre, University College London, UK*
J. A. Wiens   *Department of Biology, Colorado State University, USA*

Advisory Board
P. Adam   *University of New South Wales, Australia*
R. T. Paine   *University of Washington, Seattle, USA*
R. B. Root   *Cornell University, USA*
F. I. Woodward   *University of Sheffield, UK*

ALSO IN THE SERIES
H.G. Gauch, Jr	*Multivariate Analysis in Community Ecology*
R.H. Peters	*The Ecological Implications of Body Size*
C.S. Reynolds	*The Ecology of Freshwater Phytoplankton*
K.A. Kershaw	*Physiological Ecology of Lichens*
R.P McIntosh	*The Background of Ecology: Concepts and Theory*
A.J. Beattie	*The Evolutionary Ecology of Ant–Plant Mutualisms*
F.I. Woodward	*Climate and Plant Distribution*
J.J. Burdon	*Diseases and Plant Population Biology*
J.I. Sprent	*The Ecology of the Nitrogen Cycle*
N.G. Hairston, Sr	*Community Ecology and Salamander Guilds*
H. Stolp	*Microbial Ecology: Organisms, Habitats and Activities*
R.N. Owen-Smith	*Megaherbivores: The Influence of Large Body Size on Ecology*
J.A. Wiens	*The Ecology of Bird Communities*
N.G. Hairston, Sr	*Ecological Experiments*
R. Hengeveld	*Dynamic Biogeography*
C. Little	*The Terrestrial Invasion: An Ecophysiological Approach to the Origins of Land Animals*
P. Adam	*Saltmarsh Ecology*
M.F. Allen	*The Ecology of Mycorrhizae*
D.J. Von Wilbert *et al.*	*Life Strategies of Succulents in Deserts*
J.A. Matthews	*The Ecology of Recently-deglaciated Terrain*
E.A. Johnson	*Fire and Vegetation Dynamics*
D.H. Wise	*Spiders in Ecological Webs*
J.S. Findley	*Bats: A Community Perspective*
G.P. Malanson	*Riparian Landscapes*
S.R. Carpenter & J.F. Kitchell (Eds.)	*The Trophic Cascade in Lakes*
R.J. Whelan	*The Ecology of Fire*
R.C. Mac Nally	*Ecological Versatility and Community Ecology*

575 B439e

Evolution and Ecology: The Pace of Life

K. D. BENNETT

Department of Plant Sciences, University of Cambridge, UK

PUBLISHED BY THE PRESS SYNDICATE OF THE UNIVERSITY OF CAMBRIDGE
The Pitt Building, Trumpington Street, Cambridge CB2 1RP, United Kingdom

CAMBRIDGE UNIVERSITY PRESS
The Edinburgh Building, Cambridge CB2 2RU, United Kingdom
40 West 20th Street, New York, NY 10011-4211, USA
10 Stamford Road, Oakleigh, Melbourne 3166, Australia

© Cambridge University Press 1997

This book is in copyright. Subject to statutory exception
and to the provisions of relevant collective licensing agreements,
no reproduction of any part may take place without
the written permission of Cambridge University Press.

First published 1997

Printed in the United Kingdom at the University Press, Cambridge

Typeset in 11/13 Bembo Roman.

*A catalogue record for this book is available from
the British Library*

Library of Congress Cataloguing in Publication data

Evolution and ecology : the pace of life / K.D. Bennett
  p.   cm. – (Cambridge studies in ecology)
Includes bibliographical references. (p.   ) and index.
ISBN 0 521 39028 1 (hardback). – ISBN 0 521 39921 1 (pbk.)
1. Evolution (Biology). 2. Paleoecology – Quaternary.
I. Title. II. Series.
QH 366.2.B4635 1997
575 – dc20   96-13368 CIP

ISBN 0 521 39028 1 hardback
ISBN 0 521 39921 1 paperback

All the business of war, and indeed all the business of life, is to endeavour to find out what you don't know by what you do; that's what I called "guessing what was at the other side of the hill" *Arthur Wellesley, Duke of Wellington, 1852, quoted by J.W. Croker (Jennings 1884, vol. 3, p. 274).*

Contents

Tables	*page*	xi
Illustrations		xiii
Preface		xvii

1 Introduction — 1
    Outline — 3
    Terms and Definitions — 3

2 Development of ideas — 6
    Evolutionary processes — 6
    Ecological processes — 35
    Quaternary research — 39
    Discussion — 42

3 Orbital-forcing of climatic oscillations — 44
    Orbital parameters — 45
    Insolation — 48
    Climate models — 51

4 Geological evidence for orbital-forcing — 65
    Cenozoic — 66
    Mesozoic — 74
    Paleozoic — 82
    Proterozoic — 84
    Continental glaciation in Earth history — 85
    Discussion — 89

5 Biological response: distribution — 92
    The physical background — 92
    The terrestrial record — 94
    The marine record — 145
    Discussion — 148

6	**Biological response: evolution**	154
	Geological time	155
	Ecological time	168
	Discussion	173
7	**Biological response: extinction**	178
	Animals	178
	Plants	181
	Discussion	183
8	**Evolution and ecology: synthesis**	184
	Biological responses	184
	Post-modern evolutionary synthesis	184
	Difficulties	195
References		199
Index		226

Tables

1.1	Simplified geological time-scale	4
2.1	Darwin's model of evolution by natural selection	21
2.2	Main modes of evolution	23
3.1	Effect of Earth–Moon distance on the periodicity of palaeoclimatic parameters	47
4.1	Earth's pre-Quaternary glacial record	88
5.1	Holocene rates of increase for tree populations	151
5.2	Interglacial rates of increase for tree populations	152
6.1	Stratigraphical ranges of deer in the Quaternary of the British Isles	156
6.2	Mean species duration in geological time	175
8.1	Expanded model of evolution	187
8.2	Temporal hierarchy of processes controlling evolutionary patterns	189

Illustrations

2.1	Stages of speciation	17
2.2	Speciation and trends in lineages	28
2.3	Species response to environmental change	32
2.4	Effect of changing environments on sympatric distributions	33
3.1	Variations in the Earth's eccentricity, obliquity, and precession of the equinoxes since 1 Ma	46
3.2	Orbital variations of Mars	48
3.3	Precession of the equinoxes	49
3.4	0–100 ka deviations of solar radiation from modern values	50
3.5	GCM-simulated climatic variables for 18 ka and the present	52
3.6	GCM-simulated climatic variables for 9 ka and the present	57
3.7	Reconstructed climates of Pangaea	61
3.8	Reconstructed climatic variations with Cretaceous continental configuration	63
3.9	Reconstructed January surface temperatures during the Cretaceous	64
4.1	Geological time-series from deep-sea sediment of the Southern Indian Ocean	68
4.2	High-resolution spectra of orbital, insolation, and geological variations in the late-Quaternary	69
4.3	Orbitally-based chronostratigraphy for the late-Quaternary	71
4.4	Composite variance spectrum for oxygen isotope datasets	73
4.5	Carbonate oscillations from South Atlantic sediments	74
4.6	Miocene pollen record from Wyoming, USA	75
4.7	Sediment properties, Cretaceous, Italy	76
4.8	Power spectra of sediment variables, Cretaceous, Italy	77
4.9	Facies of the Ladinian Latemar Limestone, Triassic, northern Italy	79
4.10	Extent of detrital cycles, Lockatong Formation, Triassic, northeastern USA	80

xiv · Illustrations

4.11	Correlations and fish distributions across Newark Basin, Triassic, northeastern USA	81
4.12	Carboniferous sea-level curve	83
4.13	Thickness of calcium sulphate layers, Permian, southern USA	84
4.14	Salt content and time-series, Ordovician–Silurian, Western Australia	86
4.15	Proterozoic shelf palaeogeography, northern Canada	87
5.1	Distribution of ice-sheets at the last glacial maximum	93
5.2	Quaternary pollen record at Valle di Castiglione, Italy	97
5.3	Quaternary pollen record at Tenaghi Philippon, Greece	98
5.4	Quaternary pollen record at Ioannina, northwest Greece	101
5.5	Late-Quaternary distribution change for European *Quercus*	104
5.6	Late-Quaternary distribution change for European *Tilia*	105
5.7	Reconstructed late-Quaternary 'vegetation' units for Europe	106
5.8	European no-analogue vegetation in Europe since 13 ka	108
5.9	Changing areal extent of tree taxa in the Holocene of the British Isles	109
5.10	Quaternary pollen record from Clear Lake, California, USA	110
5.11	Late-Quaternary spread of trees in eastern North America	112
5.12	Numerical comparison of fossil and modern pollen spectra in the late-Quaternary of northwestern Canada and Alaska, USA	116
5.13	Modern and full-glacial ranges of trees and shrubs, Nevada, USA	118
5.14	Zonation of vegetation since 20 ka, Great Basin, USA	119
5.15	Zonation of plant communities since 24 ka, Arizona, USA	120
5.16	Numerical comparison of late-Quaternary and modern plant remains, Grand Canyon, USA	121
5.17	Generalized late-glacial and present plant zonation on the Colorado Plateau, USA	122
5.18	Late-Quaternary pollen record, El Valle, Panama	123
5.19	Late-Quaternary pollen record from Guatemala	124
5.20	Late-Cenozoic pollen record from Bogotá, Colombia	125
5.21	Quaternary stratigraphic ranges of pollen types from Bogotá, Colombia	127
5.22	Late-Quaternary pollen record from Kashiru, Burundi	128
5.23	Locations of Holocene pollen records, eastern Sahara	129
5.24	Holocene pollen record at Selima, Sudan	130
5.25	Quaternary pollen and charcoal record from Lynch's Crater, northeastern Australia	131

5.26	Quaternary record of pollen and charcoal from Lake George, eastern Australia	134
5.27	Late-Quaternary pollen record from New Guinea	136
5.28	Quaternary intermingled fauna: European beetles	139
5.29	Quaternary intermingled fauna: North American beetles	140
5.30	Quaternary intermingled fauna: European land mollusca	141
5.31	Quaternary intermingled fauna: North American vertebrates	142
5.32	Quaternary intermingled fauna: North American mammals	143
5.33	Quaternary intermingled fauna: Eurasian and Australian mammals	144
5.34	Full-glacial–modern distribution changes of North Atlantic coccolithophores	146
5.35	Changing Quaternary coastlines between southeast Asia and Australia	149
6.1	Last interglacial red deer from Jersey and Great Britain	155
6.2	Modern distribution of the beetle *Helophorus aquaticus* and its races	158
6.3	Morphological stasis in ostracodes of eastern North America	160
6.4	Quaternary phyletic history of the Bermudan land-snail *Poecilozonites bermudensis*	161
6.5	Comparison of fossil and recent samples of the Bahamian land-snail *Cerion agassizi*	162
6.6	Cenozoic molluscan sequence, Turkana Basin, Kenya	163
6.7	Morphological change in Pliocene foraminiferal clades of *Globoconella*	165
6.8	Pliocene evolution of equatorial Pacific radiolarians	166
6.9	Late-Cenozoic morphological change in Pacific diatom lineage *Rhizosolenia*	167
6.10	Idealized changes in interglacial distributions and abundance of European trees	169
6.11	Darwin's finches	170
6.12	Changing abundance and morphology of the finch *Geospiza fortis*, Galápagos	172
6.13	Lyellian curve for bivalve molluscs	174
7.1	Late-Cenozoic extinction episodes for North American land mammals	179
7.2	Patterns of change for North American Plio–Pleistocene mammals	180
7.3	Late-Quaternary pollen record from Easter Island	182

Preface

In July and August 1988 I spent time in Australia, partly on a study visit to the Australian National University, and partly attending a conference in Brisbane. After a conference field trip in northern Queensland, I had some time to spare in Cairns, and, as one does, took a day-trip snorkeling on the Great Barrier Reef. As the catamaran returned into Cairns, I noticed an osprey idly flying over. I had just spent much of the day watching fishes on the reef, and I had previously seen ospreys in northern Scotland, coastal Maine, and in the North American Great Lakes area (including, memorably, one flying over Exhibition Stadium, Toronto, during a ball game). I had long known that ospreys were cosmopolitan. The Cairns osprey reminded me of all of this. Whatever that bird was feeding on, whatever other organisms it interacted with, it had a different biotic environment, at least, from the Scottish or Canadian birds. I, like the rest of my generation, was brought up scientifically on the Neo-Darwinian paradigm, and had not thought too much about it in my day-to-day activities. But what, if anything, are ospreys 'adapted' to? This book is not about ospreys or their evolutionary history, but the Cairns bird has remained in my mind as a symbol for the relationships between organisms and their environments on ecological through evolutionary time-scales, including the crucial intermediate time-scales (10^4–10^5 years) typified by the Quaternary (the last 1.6 Myr).

The relationship between palaeoecology and ecology has been a topic of concern to many Quaternary workers for some time. This book was written partly as a result of a feeling of frustration that much of the ecological excitement of Quaternary events was not penetrating into mainstream ecological thinking, and partly as a result of realizing that these same events held considerable significance for processes of evolution, also untapped. The original scheme was for a book written jointly with Donald Walker on the subject of time in ecology. It evolved into this solo effort after a few years of little joint progress during which I developed the central theme of the book as it now is. Donald has been a significant source of support, ideas, and encouragement, particularly when I visited Canberra in 1988, and during the writing. His comments on an early draft convinced me to complete the thing. I am deeply indebted to him.

I began palaeoecological research under the supervision of John Birks, and I am very grateful to him for his continuing encouragement and discussion of the topics presented here, and many others, during and since those days. Many of the ideas in this book were initiated when I was an NSERC postdoctoral fellow working with Jim Ritchie while we were both in Toronto, and I thank him for his help and encouragement then and subsequently. It is an especial pleasure to thank Kathy Willis for her positive and helpful comments on several drafts, and for the many free and frank discussions that helped to shape the book. Janice Fuller allowed me to cite unpublished data from her thesis (Table 5.1) and, together with Jane Bunting, Alex Chepstow-Lusty, Simon Haberle, Susie Lumley, Maria Fernanda Sánchez Goñi, Julian Szeicz, Rebecca Teed, and Chronis Tzedakis, provided helpful discussions and made many suggestions for improvement. I also thank Nick Butterfield and Jim Ritchie for comments on a later draft, and Sylvia Peglar for allowing me to use Fig. 5.1. At CUP, I am grateful to Alan Crowden for his interest in the project, Barnaby Willitts for guiding it through the Press, and Sharon Erzinçlioğlu for carefully checking the text. John Birks and John Wiens made many helpful comments on behalf of CUP. Finally, but not least, I am grateful to Alison, Graham, and Hugh for their patience in the face of considerable neglect. If there is any merit in the book, they should all share the credit, but I alone take responsibility for omissions (it became impossible to include everything) and other failings.

The preparation of the text was greatly facilitated, and even made enjoyable, by Donald E. Knuth's wonderful type-setting program TeX, together with the macro package LaTeX (by Leslie Lamport) and the bibliography program BibTeX (by Oren Patashnik). I also gratefully acknowledge the provision of computing resources in the Department of Plant Sciences, University of Cambridge.

Keith Bennett
Cambridge
24 April 1996

1 · *Introduction*

The Earth has experienced repeated 'ice-ages', when glaciers extended over the continents and sea-levels dropped, changing the configuration of land and sea. The most recent ice-age saw the emergence of our own species, and a host of wonderful animals roamed the land. This ice-age has been an important focus of historical explanation for the origins of our modern landscapes, vegetation, and distribution of flora and fauna.

But there is much more to ice-ages than historical story-telling. Events at any period in Earth history are controlled by processes operating on a variety of time-scales, continuously from those we can see and experience directly up to the age of the Earth itself. But this range of time-scales has become divided by different academic traditions. Ecologists deal with processes operating on human time-scales and work with modern species, Quaternary palaeoecologists work with the fossil record of the same species during the period of the most recent ice-ages, and palaeontologists (mostly) occupy the rest of geological time. That is the theory. In practice, there is some dialogue between ecologists and palaeontologists, perhaps through the continuing influence of *On the Origin of Species* (Darwin 1859), but Quaternary palaeoecology receives barely token attention from either group. For example, a recent prominent symposium volume on *Perspectives in Ecological Theory* includes an 'Ecology and Evolution' section, with papers on population genetics, palaeontology, and a discussion of them (Feldman 1989; Stanley 1989; Travis & Mueller 1989). Why is the record of events on intermediate (Quaternary) time-scales not contributing to ecological and evolutionary theory? Is it because there is nothing of significance happening on those time-scales? Or because no-one is aware of what is happening? Why do Quaternary palaeoecologists keep themselves apart from these other related disciplines?

The ice-ages of the last two million years or so, provide the most accessible portion of the geological time-scale. Deposits are abundant, everywhere. Their temporal proximity means that we can work within them at time-scales measured in thousands of years, and even hundreds of years since the last major ice advance, 15,000 years ago. This temporal resolution is not normally available in the geological column. We are accustomed to the aspect of the ice-ages that treats them as part of our past, a distinctive and interesting portion of Earth history. We have been less comfortable with using the processes operating on time-scales within the ice-ages as data for our understanding of the rest of Earth history on those same time-scales. The fact that an ice-age seems like an unusual episode in Earth history has inhibited incorporation of the record in the hierarchy of time-scales influencing life on Earth.

A key point that needs to be resolved is the extent to which the Quaternary biological and environmental record can be taken as typical of the whole fossil record on the 20–100-kyr time-scale. Without it, we have no data for this time-scale, and would have to assume that processes operating in ecological time (up to hundreds of years) are effective up to geological time-scales (Gould 1985). We do have a record, but is it representative of the 20–100-kyr time-scale over the history of the Earth? What does the record indicate about evolutionary processes? To what extent is evolutionary change driven by changes in the physical environment rather than biological interactions?

This book explores answers to these questions in an attempt to address the fundamental issue of 'the business of life': how it evolves. Explicitly, the aim is to:

(i) Demonstrate that the Earth is and always has been subject to orbital variations that cause climatic changes at the surface of the Earth on Milankovitch time-scales (10^4–10^5 yr).
(ii) Determine the relative frequencies of (a) distribution change, (b) evolutionary change, and (c) extinction, where these can be shown to have occurred in clear temporal relation to climatic changes due to orbital variations.
(iii) Relate the time-scale of orbital variations to biological time-scales (life-spans of organisms, species durations).
(iv) Discuss the significance of the conclusions from aims 1–3 for theories about (a) the way in which life has evolved on Earth; and (b) the approach of ecology to interactions between organisms, and between organisms and environment.

Additionally, the history of attitudes to geological and biological data from the Quaternary will be considered in order to explore the background of the modern failure to include these time-scales in thinking about the processes that brought about the evolution of life on Earth. However, consideration of the genetical aspects of evolution is beyond the scope of this book.

This book, therefore, examines the extent to which the position of Quaternary palaeoecology at a central temporal location between ecology and palaeontology can contribute and illuminate both by extending the range of time-scales each covers. The aim is to achieve a continuum of thinking along a full range of time-scales and show that processes visible in the Quaternary record have a substantial contribution to make to ecological and evolutionary thinking. We now have the means to examine evolutionary theories against a fine geological time-scale: what is the result?

Outline

Chapter 2 covers the theoretical background of evolutionary and ecological processes. This is done through an historical approach to show the way that ideas have developed, and the state of scientific understanding about Quaternary climatic events at the time that ideas were developed. Chapter 3 describes the astronomical background for the Earth's orbital variations, how climatic change is forced by these, and then Chapter 4 presents the geological evidence for astronomically-forced climatic change throughout Earth history. In principle, the biological response to these climatic oscillations might take in any of three modes: distribution change, evolutionary change, or extinction. It is likely that all three occur, so the principal concern is with establishing the relative frequency of each mode rather than whether it exists. Response by distribution change is discussed in Chapter 5, by evolution in Chapter 6, and by extinction in Chapter 7. A post-modern synthesis of evolution and ecology across a full range of time-scales, from ecological moments to geological time, is presented in Chapter 8.

Terms and Definitions

The names and ages of stratigraphic units used follow Harland *et al.* (1990), except where there is indication to the contrary. The units 'Ga', 'Ma' and 'ka' are used to refer to dates before present in billions (10^9), millions, and thousands of years, respectively, and 'Gyr', 'Myr' and 'kyr' are used to refer to duration of time in billions, millions, and thousands of years, respectively. The term 'BP' is restricted to mean radiocarbon

Table 1.1. *Simplified geological time-scale*

Eon	Era	Period	Epoch	Age at base (Ma)
Phanerozoic	Cenozoic	Quaternary	Holocene	0.01
			Pleistocene	1.64
		Tertiary	Pliocene	5.2
			Miocene	23.3
			Oligocene	35.4
			Eocene	56.5
			Paleocene	65
	Mesozoic	Cretaceous		146
		Jurassic		208
		Triassic		245
	Paleozoic	Permian		290
		Carboniferous		363
		Devonian		409
		Silurian		439
		Ordovician		510
		Cambrian		570
Proterozoic				2500
Archaean				4000
Priscoan				4500

Source: From Harland *et al.* (1990).

years before present, where 'present' means 1950 AD. The measurement of time in radiocarbon years is not quite the same as calendar years, and calibrations have been proposed (Stuiver & Reimer 1993). Here, radiocarbon years are maintained because many of the ages concerned have entered common usage, and calibration makes little difference for the points being made.

A summary geological time-scale is given in Table 1.1 to place the geological terms used in a temporal framework. The term 'Milankovitch time-scales' is used to refer to time-scales of 10^4–10^5 yr. Throughout, the period name 'Quaternary' is used in preference to the epoch name 'Pleistocene', except in quotations, because, by covering 'Pleistocene' plus 'Holocene', it gives some coherence to a period of time that has a unity by virtue of repeated expansion and contraction of continental ice-sheets that we have every reason to believe is continuing. The term 'Holocene' is used for the most recent 10 kyr as an available name for that period of time. As yet, there are no other formally recognized subdivisions of the Pleistocene (Harland *et al.* 1990), although the deep-sea sediment stage numbers (Shackleton & Opdyke 1973), based on analyses of oxygen isotopic ratios ($\delta^{18}O$), are frequently used informally. The term 'last interglacial' is used to refer to the phase near 125 ka when the Earth's climate was last similar to the present (probably slightly warmer), and the term 'last glacial' is used to refer to the latest Quaternary cold oscillation (about 115 ka until 10 ka). Viewing the Holocene as a period of time in Earth history equivalent in status to the Pleistocene (as in 'the end of the Pleistocene') has had a stultifying effect on consideration of Quaternary events: the acceptance of the Holocene as just another phase in (so far) 1.6 Myr of Quaternary climatic oscillations cannot come soon enough (Gould 1991).

Taxonomic nomenclature of organisms should be assumed to follow the usage of the author(s) of works cited. Each scientific name used is included in the Index, together with a brief description derived from the sources used, and additional information from Honacki et al. (1982), Mabberley (1987), Carroll (1988), and Vaught (1989).

I avoid the term 'cycle' when referring to anything operating through time, because time does not 'cycle'. Instead, I use terms such as 'period' or 'oscillation' to mean quantities that vary more or less rhythmically with time.

2 · Development of ideas

This chapter presents the current status of thinking on evolutionary processes in ecological and geological time. Subsequent chapters examine actual data from the fossil record, but, since the way we see data is shaped by the ideas available for interpretation, the development of evolutionary and ecological theory will be presented first. Evolutionary theory has, unfortunately, become replete with claim and counter-claim about evolutionary processes, and the degree to which they can or cannot be resolved with the ideas of Darwin (1859). It is now impossible to obtain a balanced view by summarizing the literature, because too high a proportion of it is already secondary. Some of the primary sources went through several editions and the author's ideas shifted fundamentally, so it can make a difference which edition is being cited (Peckham 1959). If ever an area of scientific literature had an excessive ratio of words to ideas, this is it. Much of the debate has become confused by the similarity in language and terminology of various aspects of evolution, and the scientific environment at the time of writing has not received corresponding attention.

This chapter discusses the development of ideas in evolution and ecology relative to advancing knowledge of the details of Earth history, especially with respect to the Earth's orbital variations and Quaternary ice-ages. The aim is to show how the present state of affairs came about.

Evolutionary processes
Scientific beginnings

There is a long history of evolutionary and ecological thought over many centuries (Mayr 1982). But it is during the last century that those ideas that came to shape modern thinking were expressed scientifically, especially following Charles Darwin and his contemporaries, notably Charles Lyell and Alfred Wallace.

Geology and the nature of species
Charles Lyell (1797–1875) was a British geologist who dominated geology through most of the nineteenth century. His classic text *Principles of Geology* (Lyell 1830, 1832, 1833) described the geological record in terms of modern processes, and is one of the foundations of modern geology. Lyell's philosophy on the nature of species through time was covered in the second volume of the work (Lyell 1832). He began his argument by inquiring:

first, whether species have a real and permanent existence in nature; or whether they are capable, as some naturalists pretend, of being indefinitely modified in the course of a long series of generations? Secondly, whether, if species have a real existence, the individuals composing them have been derived originally from many similar stocks, or each from one only, the descendants of which have spread themselves gradually from a particular point over the habitable lands and waters? Thirdly, how far the duration of each species of animal and plant is limited by its dependance [sic] on certain fluctuating and temporary conditions in the state of the animate and inanimate world? Fourthly, whether there be proofs of the successive extermination of species in the ordinary course of nature, and whether there be any reason for conjecturing that new animals and plants are created from time to time, to supply their place? *(Lyell 1832, pp. 1–2).*

These are thoroughly modern and reasonable questions. Lyell concluded that species do have a real existence in nature, and when each first appeared, it had the attributes and organization that it does now. He thought that each species originated as a single pair (or individual, for species where that was sufficient), and that species might have been created in such a way as to multiply and spread for a particular period of time and space. He saw no evidence for the transmutation of one species into another, because there were other processes in operation that were so much more active in their nature, that they would intervene and prevent the accomplishment of any transmutation. Thus, to Lyell, the question of transmutation of species was a side issue, because it would never be likely to happen in practice even if it was, theoretically, a possibility. In answer to his own fourth question, Lyell argued that animal and plant species are dependent on certain physical conditions and on the numbers and characteristics of other species in the same region. However, all these conditions fluctuate, as a result of changing physical environments and as a result of changing distributions of species. It follows that species existing at any one period must become successively extinct. Local changes restrict the range of some species, and enable the enlargement of the ranges of others. Since new species originate in a single spot, each needs time to spread over a wide area. So there must be, simultaneously, species of recent origin and species of 'high antiquity'. He suggested, as a speculative figure, that there might be one new species and one extinction, globally, per year, which is low enough that neither extinction nor origin is likely to be observed in any one area on human time-scales (Lyell 1832).

Lyell's view on species and their origin was thus that they are created at a single spot, multiply, and spread, surviving certain environmental and biotic fluctuations, but without being transformed, and they eventually become extinct. Species, he believed, were stable units that came into existence at ecologically-appropriate points in space and time, survived for a longer or shorter period of time in dynamic ecological equilibrium with other organisms and spread to some degree, but were eventually eliminated by the pressures of the ever-changing physical and biotic environment (Rudwick 1990). Lyell was not specific about how species came into being, but his studies of contemporary geological processes showed him that the surface of the Earth is in a state of constant change, so associations of plants and animals could not always have been maintained in exactly the same spots, but shifted continually over land and sea (Worster 1985). He saw that the process of shifting distributions was likely to be a more rapid response to environmental change than *in situ* transformation, because there would always be species around better suited to new conditions than the species on the spot originally. He pointed out that, following climatic change, some species would be preserved by shifting their distributions, but the same change would be 'fatal to many which can find no place of retreat, when their original habitations become unfit for them' (Lyell 1832, p. 170). He noted that:

if a tract of salt-water becomes fresh by passing through every intermediate degree of brackishness, still the marine molluscs will never be permitted to be gradually metamorphosed into fluviatile species; because long before any such transformation can take place by slow and insensible degrees, other tribes, which delight in brackish or fresh-water, will avail themselves of the change in the fluid, and will, each in their turn, monopolize the space *(Lyell 1832, p. 174)*.

His views on the fixity of species, and their occurrence under certain conditions of the physical and biotic environment lead to his notorious declaration that, under the right circumstances:

the huge iguanadon might reappear in the woods, and the ichthyosaur in the sea, while the pterodactyle might flit again through umbrageous groves of tree-ferns *(Lyell 1830, p. 123)*.

This seems a ludicrous statement today, and surprising from such an eminent and respected geologist. But it does follow, with a certain devastating logic, from Lyell's views on the cycling of global environments and the origin of species. This passage and its interpretation in the light of Lyell's views on geological time have been analyzed by Gould (1987). Much of the underlying thinking of Lyell on species, however, remains valid despite the spectacular conclusion, and the end point should not be allowed to obscure the basis of the arguments.

In all, there were 12 editions of *Principles of Geology* and the later editions incorporated, amongst other new material, two ideas not available to Lyell in the 1830s. The first was a discussion of Darwin's ideas about evolution and natural selection, and the second was the notion that global climatic change is forced by variations in the Earth's orbital parameters.

Lyell learned of Darwin's ideas about evolution and natural selection before the publication of *On the Origin of Species* (Darwin 1859), and was not convinced, initially. Lyell worked extensively with Tertiary molluscs, many of which can be identified to living species, and he appreciated that they must have survived changing climates since the Tertiary (Lyell 1830, 1832, 1833), which appeared to conflict with Darwin's ideas. In a letter to Darwin dated 17 June 1856, Lyell wrote:

And why do the shells which are the same as European or African species remain quite unaltered like the Crag species which returned unchanged to the British seas after being expelled from them by Glacial cold, when 2 millions? of years had elapsed, and after such migration to milder seas. Be so good as to explain all this in your next letter *(Burkhardt & Smith 1990, p. 146)*.

This question came at the end of a long letter about uplift of continents, and Darwin commented on that aspect of the letter (in a postscript on 18 June 1856 of a letter to Joseph Hooker, and in reply to Lyell on 25 June 1856 (Burkhardt & Smith 1990, p. 147, pp. 153–155). However, it appears that Darwin did not reply directly to Lyell's closing question, or comment on it anywhere else. Lyell copied the main points of his question into his journal (Wilson 1970, pp. 104–105), which suggests that he thought it was important (Burkhardt & Smith 1990).

Lyell eventually shifted his position on the question of species, accepting the idea of gradual change, but he did notice and comment again on the paradox of little change through 'the Glacial period':

we have yet found evidence that most of the testacea, and not a few of the quadrupeds, which preceded, were of the same species as those which followed the extreme cold. To whatever local disturbances this cold may have given rise in the distribution of species, it seems to have done little in effecting their annihilation. We may conclude therefore, from a survey of the tertiary and modern strata, which constitute a more complete and unbroken series than rocks of older date, that the extinction and creation of species have been, and are, the result of a slow and gradual change in the organic world *(Lyell 1875, vol. 1, pp. 306–307).*

The idea that global climatic change is forced by varying orbital parameters had been around earlier, but only received serious attention after the mathematics were worked out by the Scottish geologist and philosopher James Croll, and published prominently in the scientific literature (Croll 1864, 1865, 1866, 1867a, b, 1868, 1875). Lyell covered this at length in the final edition of *Principles of Geology*, appreciating that there had been more than one glacial phase since the Tertiary (Lyell 1875). However, he did not realize that there had also been pre-Quaternary glaciations, and he seems to have considered that Quaternary events had little relevance for Earth history in general.

Evolution by means of natural selection
The theory of evolution by means of natural selection was developed independently by two naturalists, Alfred Wallace and Charles Darwin. Darwin was the older of the two, and worked his ideas out first. He discussed them with friends, but did not publish them until receipt of a manuscript from Wallace along similar lines forced the matter into the public domain.

Alfred Wallace (1823–1913) was an English naturalist and explorer, who spent much of his life working in the rain forests of southeast Asia and the Amazon basin. His contribution, initially as an essay sent to Darwin in 1858, was read at a meeting of the Linnean Society, London, together with documents written by Darwin, demonstrating that they had developed similar views independently. The details of how this arrangement came to be set up can be found in a paper by Beddall (1988), and references therein.

Wallace began by establishing two points, via discussion of the tendency of populations to increase geometrically if left unchecked:

1st, that the animal population of a country is generally stationary, being kept down by a periodical deficiency of food, and other checks; and 2nd, that the comparative abundance or scarcity of the individuals of the several species is entirely due to their organization and resulting habits, which, rendering it more difficult to procure a regular supply of food and to provide for their personal safety in some cases than in others, can only be balanced by a difference in the population which have to exist in a given area *(Wallace 1859, p. 57)*.

He then noted the existence of varieties, and that most changes would affect survival ability, one way or the other. Next:

let some alteration of physical conditions occur in the district — a long period of drought, a destruction of the vegetation by locusts, the irruption of some new carnivorous animal seeking "pastures new" — any change in fact tending to render existence more difficult to the species in question, and tasking its utmost powers to avoid complete extermination; it is evident that, of all the individuals composing the species, those forming the least numerous and most feebly organized variety would suffer first, and, were the pressure severe, must soon become extinct. The same causes continuing in action, the parent species would next suffer, would gradually diminish in numbers, and with a recurrence of similar unfavourable conditions might also become extinct. The superior variety would then alone remain, and on a return to favourable circumstances would rapidly increase in numbers and occupy the place of the extinct species and variety *(Wallace 1859, p. 58)*.

In due course, new varieties appear, the process repeats, and we get progression and continued divergence deduced from the general laws which regulate the existence of animals in a state of nature, and from the undisputed fact that varieties do frequently occur *(Wallace 1859, p. 59)*.

This progression, by minute steps, in various directions, but always checked and balanced by the necessary conditions, subject to which alone existence can be preserved, may, it is believed, be followed out so as to agree with all the phenomena presented by organized beings, their extinction and succession in past ages, and all the extraordinary modifications of form, instinct, and habits which they exhibit *(Wallace 1859, p. 62)*.

The essay was too brief for Wallace to develop arguments as fully as Darwin did in *On the Origin of Species*, written in the immediate aftermath of the joint presentation at the Linnean Society. It parallels Darwin's longer argument in some respects, but differs in others. In particular, Wallace was explicit about populations being generally stationary, and about selection taking effect when a population is stressed ('some alteration of physical conditions'). There are hints here that the two hypotheses might have differed more substantially if Wallace had ever developed his ideas fully.

Charles Darwin (1809–1882) was another English scientist who, after an early period travelling round the world as a naturalist on the *Beagle*, a survey ship, spent the rest of his life working and thinking about a wide range of biological and geological problems, including the first full development of a coherent theory of evolution that included the principle and a plausible explanation. This work (Darwin 1859) is, of course, generally seen as the defining document of evolution. However, far more has been written and claimed about what Darwin said than what he actually did say. What were the arguments he presented? He established the principle of descent with modification, and also the principle that this is accomplished by means of a process that he termed 'natural selection', to distinguish it from 'artificial selection'. The principle of descent is now generally accepted, and is not discussed further here. But the nature of the process is a matter of active debate. Darwin presented his argument for the existence of natural selection in summary and at length several times, for example:

If during the long course of ages and under varying conditions of life, organic beings vary at all in the several parts of their organisation, and I think this cannot be disputed; if there be, owing to the high geometrical powers of increase of each species, at some age, season, or year, a severe struggle for life, and this certainly cannot be disputed; then, considering the infinite complexity of the relations of all organic beings to each other and to their conditions of existence, causing an infinite diversity in structure, constitution, and habits, to be advantageous to them, I think it would be a most extraordinary fact if no variation ever had occurred useful to each being's own welfare, in the same way as so many variations have occurred useful to man. But if variations useful to any organic being do occur, assuredly individuals thus characterised will have the best chance of being preserved in the struggle for life; and from the strong principle of inheritance they will tend to produce offspring similarly characterised. This principle of preservation, I have called, for the sake of brevity, Natural Selection *(Darwin 1859, pp. 126–127)*.

I do believe that natural selection will always act very slowly, often only at long intervals of time, and generally on only a very few of the inhabitants of the same region at the same time. I further believe, that this very slow, intermittent action of natural selection accords perfectly well with what geology tells us of the rate and manner at which the inhabitants of this world have changed *(Darwin 1859, pp. 108–109)*.

Darwin thus presented natural selection as the conclusion of a strong deductive argument: if certain conditions hold, then natural selection is the result. The point has been emphasized, and the structure of the argument formalized by Huxley (1942) and Mayr (1993) (see page 20), among others, and treated as an algorithm by Dennett (1995). Darwin's argument dealt immediately with species as temporary stages in continual, gradual change as a consequence of increasing minute adaptation to other species and the environment (Gould 1993a), making ecological interactions the driving force for macroevolution, and hence giving them a dominant role in the course of the history of life on Earth.

He suggested that acquired characters could be inherited, and that natural selection could operate with this source of variation, concluding that use and disuse have often modified the structure and functioning of organs, but that the effects are combined with or, sometimes, lost by the operation of natural selection on innate differences (Darwin 1859, pp. 142–143). However, ignorance of the mechanism of variation in the mid-nineteenth century was total. Darwin simply adopted a 'black box' approach to this difficulty: it didn't matter how offspring came to be distinct from their parents, as long as there were such differences, and natural selection could operate on them.

Darwin then turned to the geological record, and devoted much space to arguing that it is an imperfect record. But, given that it is imperfect, then natural selection means that all forms of life are connected by generation into in a single 'grand system' (Darwin 1859, p. 344), and much then becomes explicable. It is significant that Darwin argued first of all for the existence of natural selection, and then that this meant that all forms of life are connected by common descent. It would have been more natural to reverse the argument: first, common descent, and then a second argument, the process that brought this about. He appears to have decided that the geological record was too weak to make the case for common descent on its own, so it had to be made via the process.

On the basis of his theory of natural selection, Darwin stated that we can clearly understand why a species when once lost should never reappear, even if the very same conditions of life, organic and inorganic, should recur (Darwin 1859, p. 315), which is as clear a refutation of Lyell (1830, p. 123, quoted on page 9) as one could get.

In his Chapter 11, Darwin began discussion of geographical distribution, and included an extended commentary (18 pages) on the effects of 'the glacial period'. He suggested that plants and animals of the northern hemisphere would have generally spread southwards during time of increasing cold, and then spread back northwards during the following climatic warming:

The arctic forms, during their long southern migration and re-migration northward, will have been exposed to nearly the same climate, and, as is especially to be noticed, they will have kept in a body together; consequently their mutual relations will not have been much disturbed, and, in accordance with the principles inculcated in this volume, they will not have been liable to much modification *(Darwin 1859, p. 368).*

This line of argument became more explicit in later editions of *On the Origin of Species*, following publication of calculations by James Croll (1864, 1865, 1866, 1867a, b, 1868, 1875) of the Earth's orbital variations. Croll predicted that periods of glaciation would be out of phase between the north and south hemispheres, with the hemisphere angled towards the Sun at perihelion (see Chapter 3) being warm, and the other cold. Darwin argued that this provides a mechanism for the spread of cold-tolerant species from one hemisphere to the other:

Thus, we should have some few species identically the same in the northern and southern temperate zones and on the mountains of the intermediate tropical regions. But the species left during a long time on these mountains, or in opposite hemispheres, would have to compete with many new forms and they would be exposed to somewhat different physical conditions; hence they would be eminently liable to modification, and would generally now exist as varieties or as representative species; and this is the case *(Darwin 1872, pp. 339–340).*

Again, we have Darwin's view that modifications will occur when the environments of species are altered by changes such as those related to continental glaciation, although it should be noted that we now know that the northern and southern hemispheres are not out of phase with respect to periods of glaciation (see page 70).

Early twentieth century

The chief contribution of scientists working at the turn of the century, from the perspective of later generations of evolutionary biologists, was the discovery of the principles of heredity and the development of genetics. One of the leaders in this field at the time was the Dutch plant physiologist Hugo de Vries. He was one of the co-discoverers of Mendel's work, and developed his own theory of evolution based on the origin of species by mutation. Subsequent generations have been dismissive of de Vries' work, rejecting the origin of species by mutation as atypical, or based on incomplete knowledge of the complex genetic system of the plant, *Oenothera*, with which he worked (Dobzhansky 1937; Mayr 1942; Simpson 1944; Stebbins 1950). However, Gould (1993b) has pointed out that de Vries (1906) produced a coherent theory of evolution that is strikingly convergent on the punctuated equilibria hypothesis (Eldredge & Gould 1972), in all respects except the precise means of rapid speciation: mutation for de Vries (1906), geographic isolation for Eldredge & Gould (1972).

The views of de Vries (1906) were generally ignored and have had to be rediscovered by subsequent generations. They are relevant here because de Vries was aware of the problem of species survival through the last glacial period, but his conclusion from similar data contrasts with that of Darwin (1859, p. 368, quoted on pages 13–14). De Vries, like Darwin, noticed the occurrence of widely separated populations of arctic and alpine species at scattered localities in high latitudes and at high altitudes:

As no transportation over such large distances can have brought them from one locality to the other, no other explanation is left than that they have been wholly constant and unchanged ever since the glacial period which separated them. Obviously they must have been subjected to widely changing conditions. The fact of their stability through all these outward changes is the best proof that the ordinary external conditions do not necessarily have an influence on specific evolution *(de Vries 1906, p. 696)*.

De Vries (1906), therefore, believed that species could not maintain their mutual relations during changing climates associated with a glacial period, and he thought, like Lyell (see page 9), that species would persist unchanged through changing glacial climates, whereas Darwin (1859) took the opposite view (see page 13). For both, of course, this was a matter of belief, since there were no data then available to establish the point either way.

Evolutionary 'synthesis'

A single decade saw the appearance of a series of books that have been acknowledged as cornerstones of a 'modern synthesis' of evolutionary theory. Three (Dobzhansky 1937; Mayr 1942; Simpson 1944) are central to the amalgamation of the new science of mathematically-based population genetics (Fisher 1930; Wright 1931; Haldane 1932) with Darwinian evolution as practised in animal systematics and palaeontology, while Huxley (1942) gave the approach a name, and Stebbins (1950) extended it to plant systematics. The core of the synthesis is the acceptance of two conclusions:

Gradual evolution can be explained in terms of small genetic changes ("mutations") and recombination, and the ordering of this genetic variation by natural selection; and the observed evolutionary phenomena, particularly macroevolutionary processes and speciation, can be explained in a manner that is consistent with the known genetic mechanisms *(Mayr 1980, p. 1)*.

However, there are many intricacies in the way that the modern synthesis became established. These are developed in the following sections. A more detailed review is provided by Eldredge (1985).

Genetics

Theodosius Dobzhansky (1900–1975) was a Russian-born American geneticist. His book, *Genetics and the Origin of Species* (Dobzhansky 1937), showed how the relatively new knowledge of genetics could be combined with a theory of evolution. Much of the work is, in fact, a genetics textbook, but written from an evolutionary point of view. Dobzhansky contrasted, explicitly, two views of inheritance. Under the blending view, new variability inevitably becomes diluted over successive generations, unless it is renewed *de novo* at a rate that at least exceeds its rate of loss. If inheritance is particulate, as shown by Mendel, there is no such loss. Obsolescence of the blending theory following the rediscovery of Mendel's work, which was unknown to Darwin, removed what had been one of the most significant obstacles to the development of evolutionary theory, and Dobzhansky was at pains to point this out (Dobzhansky 1937). Dobzhansky was clear that natural selection is to do with 'adaptation', but was open on whether natural selection (the process) could also explain evolution (the fact), hinting that there was no agreement on this point. His view on the response of populations to environmental change was that gene frequencies would fluctuate (Dobzhansky 1937, p. 179). Later in the same chapter, Dobzhansky discussed the effect of population size, pointing out that a population split into isolated colonies might well become differentiated, even under uniform environments, and that such change need not be adaptive.

Dobzhansky (1937) did not cite *On the Origin of Species* at all, although mention of Darwin is indexed for ten pages. 'Natural selection' is not indexed, but 'selection' is. The book established contemporary theory of the evolutionary process, and backward glances to the history of the idea are minimal. All of the genetic basis of Dobzhansky's argument was unknown to Darwin, so Dobzhansky may not have seen Darwin's argument for natural selection as being particularly relevant.

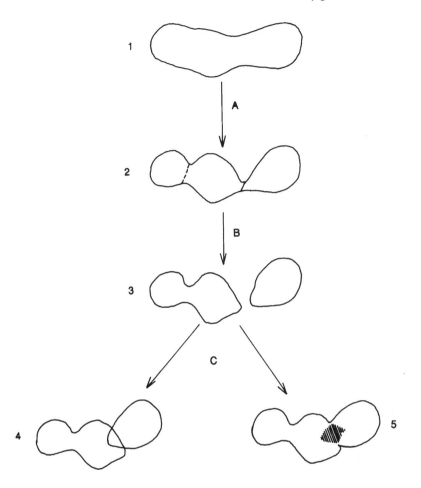

Figure 2.1 Stages of speciation. A uniform species with a large range (stage 1) differentiates into subspecies (process A). This results in a geographically-variable species with a more or less continuous array of similar subspecies (stage 2). The subspecies then become isolated by either the action of geographic barriers between populations, or development of isolating mechanisms in the isolated and differentiating subspecies (process B). The species becomes geographically variable with many subspecies completely isolated, especially near the border of the range, and some of them morphologically as different as good species (stage 3). Isolated populations then expand their ranges into the territory of the representative forms (process C). This results in either non-crossing, that is, new species with restricted range (stage 4), or interbreeding, that is, the establishment of a hybrid zone (stage 5). Redrawn from Mayr (1942, Fig. 16).

Systematics and geographic speciation

Ernst Mayr, a zoological systematist and philosopher of biology, began his career working on the bird systematics of Pacific islands. His contribution to the developing evolutionary synthesis began with *Systematics and the Origin of Species* (Mayr 1942), which includes his famous 'biological' species definition: 'species are groups of actually or potentially interbreeding natural populations, which are reproductively isolated from other such groups' (Mayr 1942, p. 120). However, he saw species as intergrading continuously through time, and argued strongly for the importance of geographic speciation (Fig. 2.1), although the process would be a slow and continuous one (Mayr 1942, p. 159).

Mayr (1942) was keen to point out that geographic isolation is, for most groups of organisms, one of the necessary conditions for speciation, although he did also detail the possibilities for non-geographic speciation, termed 'sympatric' speciation, to distinguish it from 'allopatric', or geographic speciation. He was careful to draw out the differences between his views and those of Darwin, emphasizing especially population-thinking, as a corollary to geographic speciation, in contrast to Darwin's emphasis on individuals and competition. He discussed evidence for the rate of speciation, presenting various lines of argument, none of which he found particularly convincing, but his examples indicate that he was thinking in terms of thousands of years, that speciation is more rapid in small, isolated populations, and that these rates are 'not too slow to account for the present multitude and diversity of life'.

In an interesting observation on the effect of Quaternary climatic oscillations, Mayr suggested that glaciations may be influential in species formation:

Climatic events, such as the deterioration of the climate during the Pleistocene glaciation, or the drying up of parts of the tropical continental shelves during the height of the glaciation, or the opening up of vast continental areas behind the retreating ice during the post-Pleistocene era, have all contributed toward the shifting of populations. As far as the process of speciation is concerned, these periods of range expansion have primarily two effects. First, they lead to an expansion of the restricted ranges of isolated species, a process often accompanied by the jumping of minor geographic barriers, and set the stage for the beginning of the process of diversification, which is the beginning of all speciation. Secondly, they produce an overlap of ranges among numbers of closely allied and formerly strictly allopatric species. In other words, they complete the process of speciation. Range expansions are thus of vital importance, in connection with both the beginning and the conclusion of the speciation process *(Mayr 1942, pp. 239–240)*.

A chapter on the biology of speciation dominates Mayr's book. He discussed factors influencing speciation in a complex and detailed hierarchy, but omitted any consideration of the relative importance of the various possible factors. A reference back to one of the chapters on geographic variation reveals, however, that he, like Dobzhansky, did not assume that all change was adaptive. Mayr (1942) was very clear that the origin of higher categories was nothing more than an extrapolation of speciation, which itself can be traced back to infraspecific variation.

Mayr (1942) acknowledged that his work built on the foundation in genetics laid by Dobzhansky (1937), and the number of index entries for Dobzhansky exceeds those for Darwin. *On the Origin of Species* is cited, but 'natural selection' is not indexed. 'Selection' receives four index entries, and there are a handful more for 'adaptation'. As in Dobzhansky's (1937) book, there is a strong feel of a break with the past. The book reads as if meant to be part of a new start.

Ernst Mayr has been able to maintain a high level of activity to the present day in the field of evolutionary theory. He developed a more detailed description of the allopatric speciation ideas of Mayr (1942), including a passage that anticipated the punctuated equilibria hypothesis of 20 years later (see page 26).

Rapidly evolving peripherally isolated populations may be the place of origin of many evolutionary novelties. Their isolation and comparatively small size may explain phenomena of rapid evolution and lack of documentation in the fossil record, hitherto puzzling to the paleontologist *(Mayr 1954, p. 179)*.

Mayr (1963) discussed geographic speciation in considerable depth, emphasizing again its importance relative to sympatric speciation. More significantly, for the present purpose, he expanded his views on the evolutionary role of species. He described species as biological experiments, quite comparable to mutations in their evolutionary role. Most will not persist, but a rare few will form the basis of a new evolutionary advance, with no way to predict whether a new species is an evolutionary dead-end or not. A prerequisite for evolutionary progress is a prodigious multiplication of species, because without it there would be no diversification, adaptive radiation, and little 'evolutionary progress' (Mayr 1963, p. 621).

Mayr (1963) developed the concept of geographic speciation to its limits, and made a reasonable inference about the likely macroevolutionary consequences of the concept. In so doing, he took the theory of

20 · Development of ideas

evolutionary mechanisms away from Darwin (1859) in two areas: the emphasis on isolation, and the emphasis on species. He might have gone further, but his knowledge and use of geological data were weak and even, in the case of glaciation, naïve.

In several recent syntheses of the history of evolutionary thought, Mayr (1982, 1988, 1993) has developed an exposition of evolution through natural selection based on five 'facts' and three 'inferences' (Table 2.1: see also Mayr 1982, 1988, 1993). Fact 2 is interesting for being not in the least controversial, either in Darwin's time or later (Huxley 1942; Mayr 1942, 1982, 1988, 1993). Mayr (1993, p. 86) notes that 'only a few (on average, two) of all the offspring of a set of parents survive long enough to reproduce'. His breakdown of this component of Darwin's view of natural selection is useful (see page 36).

Palaeontology

George Simpson (1902–1984) was an American palaeontologist. His book, *Tempo and Mode in Evolution* (Simpson 1944), was the last of the three core books of the modern synthesis to have been published, but had been somewhat delayed by the demands of war. Simpson, in his preface, commented that the work had been begun in 1938, and that he had been unable to revise a final version as he would have wished. There is no mention of Mayr (1942), and the two works should be considered as having been written in parallel, rather than one after the other.

Simpson (1944) was interested in time, and the significance of the palaeontological record for distinguishing between various possible modes of evolution. The opening passage of the book reads:

How fast, as a matter of fact, do animals evolve in nature? That is the fundamental observational problem of tempo in evolution. It is the first question that the geneticist asks the paleontologist *(Simpson 1944, p. 3)*.

He discussed the information that was available pertaining to rates of evolution in the first chapter, but complained that it was mostly too vague and unsatisfactory to be useful. This section of the book is way ahead of its time. Simpson had worked out the questions to ask of the fossil record, but available dating control was simply not good enough. His following chapter, on 'determinants of evolution' develops themes of variability, rate and character of mutations, length of generations, population size, and selection, and he used as an example events at 'the close of the Pleistocene' (the transition from the last glacial into the present interglacial):

Table 2.1. *Darwin's deductive model of evolution through natural selection*

Fact 1 Potential exponential increase of populations (superfecundity) *plus* **Fact 2** Observed steady-state stability of populations *plus* **Fact 3** Limitation of resources *Therefore:*	**Inference 1** Struggle for existence among individuals *plus* **Fact 4** Uniqueness of the individual *plus* **Fact 5** Heritability of much of the individual variation *Therefore:*	**Inference 2** Differential survival, i.e. natural selection *Therefore:* **Inference 3** Through many generations: evolution

Source: From Mayr (1993, Fig. 1).

a greater proportion of large animals became extinct than of small animals. Throughout the long Pleistocene epoch (a million years, more or less) these large forms, with long generations, would tend to make a slow genetic adjustment to the lower average temperature, or generally inclement climates, in a specific region, and readjustment to the more moderate recent climates would be slow. Smaller animals would adjust and readjust more rapidly and might tend to adapt primarily, not to the long average environment, but to a short summer–winter swing within the life cycle, more severe, but of the same nature as the shifts in their present life cycles *(Simpson 1944, pp. 64–65).*

Simpson (1944) commented that Fisher (1930), Wright (1931), and

Haldane (1932) agreed that evolution without selection would be slow in large populations. It might, by this process, take between 1 and 10 million years for a subspecific advance in the Equidae without natural selection. However, one lineage had evolved through nine genera in 45 million years, so some other process must be at work, and he argued that that process must be natural selection (Simpson 1944, p. 81).

He summarized his arguments on selection by reference to three theories that have denied crucial importance to selection: Lamarckian, vitalistic, and preadaptation theories, and rejected each in turn. Simpson (1944) argued that numerous examples showed that 'mega-evolution' normally occurred among small populations that become preadaptive and evolved continuously and rapidly to radically different ecological positions (Simpson 1944, p. 123). He noted that the splitting of populations into small groups was the genetical situation most likely to generate new forms.

Simpson (1944) concluded with a chapter on modes of evolution, in which he defined and described three modes: speciation, phyletic evolution, and quantum evolution. His characterizations of the main features of each type are presented here as Table 2.2.

Darwin is mentioned only once, but references to 'natural selection' and 'selection' are frequent, including index entries. Dobzhansky's work (including Dobzhansky 1937) is mentioned frequently.

Plants

The core books of the modern synthesis, and *Evolution: the Modern Synthesis* (Huxley 1942) itself, were all written by zoologists or from a zoological perspective. Credit for bringing plant evolution into the modern synthesis is usually accorded to Stebbins (1950) for *Variation and Evolution in Plants*, 'intended as a progress report on this synthetic approach to evolution as it applies to the plant kingdom' (Stebbins 1950, p. ix). Consciously writing a text to bring plants into the synthesis, Stebbins (1950) did not claim to add to it, or develop any fundamental new hypotheses about evolution, and the book has not subsequently received the attention of the founding works of the modern synthesis (for example by Eldredge 1985). However, Stebbins (1950) did develop a distinctive botanical view on evolution in two areas: modes of speciation that are largely confined to plants, and recognition of the potential significance of Quaternary climatic changes as an agent for causing distribution change and isolation. Stebbins (1950) also based his book heavily on Darwin and natural selection.

Table 2.2. *Characteristics of the main modes of evolution*

Mode	Speciation	Phyletic evolution	Quantum evolution
Typical taxonomic level	Low; subspecies, species, genera, etc.	Middle; genera, subfamilies, families, etc.	High; families, suborders, orders, etc.
Relation to adaptive grid	Subzonal	Zonal	Interzonal
Adaptive type	Local adaptation and random segregation	Postadaptation and secular adaptation; (little inadaptive or random change)	Preadaptation (often preceded by inadaptive change)
Direction	Shifting, often essentially reversible	Commonly linear as a broad average, or following a long shifting path	More rigidly linear, but relatively short in time
Typical pattern	Multiple branching and anastomosis	Trend with long-range modal shifts among bundles of multiple isolated strands, often forked	Sudden shift from one position to another
Stability	Series of temporary equilibria, with great flexibility in minor adjustments	Whole system shifting in essentially continuous equilibrium	Radical or relative instability with the system shifting toward an equilibrium not yet reached
Variability	May be temporarily depleted and periodically restored	Nearly constant in level; most new variants eliminated	May fluctuate greatly; new variants often rapidly fixed
Typical morphological changes	Minor intensity; colour, size, proportions, etc.	Similar to speciation, but cumulatively greater in intensity; also polyisomerisms, anisomerisms, etc.	Pronounced or radical changes in mechanical or physiological systems
Typical population involved	Usually moderate with imperfectly isolated subdivisions	Typically large isolated units, with speciation proceeding simultaneously within units	Commonly small wholly isolated units
Usual rate distribution	Erratic or comparable to horotelic rates	Bradytelic and horotelic	Tachytelic

Source: From Simpson (1944).

Stebbins (1950) began his book with accounts of the nature of variation in plants, natural selection, and genetic systems. The core of the book, however, is dominated by chapters on hybridization and polyploidy, processes that have much greater significance for evolution in plants than they do among animals. These processes, operating together or separately, greatly increase the potential for sympatric speciation, and hence reduce the relative importance of isolation and allopatric mechanisms for speciation among plants.

The study of Quaternary climatic changes was beginning to advance by 1950, but still had no reliable and accurate time-scale (see page 41). It was clear, however, that there had been repeated phases of glaciation, and Stebbins (1950) made full use of this knowledge. The book has numerous references to Quaternary events and their significance for plant distribution and evolution, and Stebbins was thus probably the first major contributor to the modern synthesis to appreciate the extent to which climatic changes might influence the course of evolution. He argued that environmental change (such as that occurring within the Quaternary) was essential for rapid evolution. However, the ability of a group to respond to change depended on biological characteristics of the group, including such features as having a high degree of genetic heterozygosity, favourable population structure, and morphological or physiological characteristics that are potentially preadaptive in the direction of the environmental change. Populations that fail to respond to environmental change become reduced in size and, eventually, extinct. The end result is that rates of evolutionary change can be rapid ('explosive' on geological time-scales), or so slow as to be 'essentially static'.

Thus, Stebbins (1950) followed the pattern of the other books of the modern synthesis, for example in his general treatment of adaptation and natural selection, while emphasizing those features of evolution that are of especial significance among plant groups. He saw the evolution of higher categories as a continuation of the processes that gave rise to subspecies and species, taking place over longer time-scales and with additional genetic and environmental change, and he supported Mayr's (1942) biological species concept (see page 18) in principle, but pointed to a number of practical difficulties that arose when applying the principle to plant groups. The differences are not greater than the similarities, but the end result is a work that is much more than modern synthesis applied to plants.

Stebbins (1984) developed the earlier ideas on the consequences of Quaternary environmental change for evolutionary change into his 'secondary contact hypotheses'. He argued that repeated Quaternary glaciations forced alpine and arctic plants to lower altitudes and latitudes, bringing previously isolated populations into contact. Hybridization and polyploidy generated new races and species. Thus some arctic and alpine species may date from the time of the last period of major ice retreat, perhaps 14,000 years ago.

Modern synthesis
Julian Huxley (1887–1975) was an influential British zoologist with a wide range of interests. His book on evolution (Huxley 1942) launches straight into a vigorously Darwinian stance, analysing Darwin's argument as follows:

> Darwin based his theory of natural selection on three observable facts of nature and two deductions from them. The first fact is the tendency of all organisms to increase in a geometrical ratio ... The second fact is that, in spite of this tendency to progressive increase, the numbers of a given species actually remain more or less constant.
> The first deduction follows. From these two facts he deduced the struggle for existence ...
> Darwin's third fact of nature was variation: all organisms vary appreciably. And the second and final deduction, which he deduced from the first deduction and the third fact, was Natural Selection *(Huxley 1942, p. 14)*.

Huxley (1942) commented that the first two facts were unquestioned. Darwin (1859) certainly discussed the first. However, the second was left implicit in his discussion of the 'checks' that prevent a population from increasing according to its potential, but was made explicit by Wallace (1859, p. 57, quoted on pages 10–11).

Huxley's work received little attention, except for giving a name to the burgeoning consensus among biologists and geologists about evolution. It differed from the works of Dobzhansky (1937), Mayr (1942), and Simpson (1944) in the prominence it gave to Darwin and to natural selection, while still covering the same genetical ground as, for example, Dobzhansky (1937). It is striking that the three main works of the synthesis paid scant attention to Darwin, whereas Huxley, who named the new synthesis, placed the structure of Darwin's argument at the centre of his book. Dobzhansky (1937), Mayr (1942), and Simpson (1944) all built their arguments on an understanding of the natural world that was not available to Darwin. Perhaps, ultimately, Huxley (1942) contributed little new: it is very much a text, while Dobzhansky (1937), Mayr (1942), and Simpson (1944) all broke new ground in some way.

Development of ideas

The modern synthesis is still prevalent, and is strenuously defended by a group of evolutionary biologists who have been described as 'ultra-Darwinians': slavish adherents to the Darwinian natural selection model of generation-by-generation change extrapolated back through geological time (Eldredge 1995), which is nothing more than substantive uniformitarianism (*sensu* Gould 1965). Stebbins (1972) has made this explicit in his hypothesis of genetic uniformitarianism, arguing that the processes of evolution (mutation, genetic recombination, natural selection, reproductive isolation) have operated in the past in the same manner as now, although genotypes, phenotypes, and environmental conditions are variable and specific to time and place. Hence, macroevolution can be studied through attention to microevolution in modern plant populations. Other arguments on the 'ultra-Darwinian' view of evolution are set out in books by George Williams (1966, 1992), John Maynard Smith (1975), Richard Dawkins (1976, 1983, 1986, 1995), and Daniel Dennett (1995).

Perhaps the most extreme view within the modern synthesis is the 'Red Queen hypothesis' (Van Valen 1973), which proposes that evolution is a zero-sum game. No matter how well species adapt to their environments, they remain in the same relative position as other species because they are all adapting as well. Extinction probabilities through time remain more or less constant because, overall, species cannot get better at surviving (Ridley 1994).

Synthesis reconsidered

Thinking on the processes of evolution since the establishment of the modern synthesis has been dominated by two developments: the philosophical question of the nature of species, and the interpretation of the fossil record as a record of stasis interrupted, for any given lineage, by relatively brief periods of rapid change.

Punctuated equilibria

Niles Eldredge and Stephen Jay Gould are American palaeontologists who set out the arguments for punctuated equilibria in a paper (Eldredge & Gould 1972) that addressed directly the question of how palaeontologists viewed speciation, arguing:

(1) The expectations of theory color perception to such a degree that new notions seldom arise from facts collected under the influence of old pictures of the world. New pictures must cast their influence before facts can be seen in different perspective.

(2) Paleontology's view of speciation has been dominated by the picture of "phyletic gradualism". It holds that new species arise from the slow and steady transformation of entire populations. Under its influence, we seek unbroken fossil series linking two forms by insensible gradation as the only complete mirror of Darwinian processes; we ascribe all breaks to imperfections in the record.

(3) The theory of allopatric (or geographic) speciation suggests a different interpretation of paleontological data. If new species arise very rapidly in small, peripherally isolated local populations, then the great expectation of insensibly graded fossil sequences is a chimera. A new species does not evolve in the area of its ancestors; it does not arise from the slow transformation of all its forbears. Many breaks in the fossil record are real.

(4) The history of life is more adequately represented by a picture of "punctuated equilibria" than by the notion of phyletic gradualism. The history of evolution is not one of stately unfolding, but a story of homeostatic equilibria, disturbed only "rarely" (i.e., rather often in the fullness of time) by rapid and episodic events of speciation *(Eldredge & Gould 1972, pp. 83–84)*.

Eldredge & Gould (1972) did not define what they meant by 'rapid', but have since indicated that time-scales of 5–50 kyr might be considered 'rapid' (Gould 1982; Eldredge 1986, 1989, 1995): slow from the perspective of geneticists, but fast on geological time-scales. The general approach produced a storm of protest (see page 34), although at heart the hypothesis is Mayr's peripatric theory of speciation translated into geological time (Gould 1994), and the key point had been recognized, but not developed, by Mayr (1954, p. 179, quoted on page 19). However, the ideas of Eldredge & Gould (1972) were addressed at palaeontologists, and they introduced a sense of temporal scale lacking in the earlier writing of Mayr (1942, 1954, 1963) and others. Point (2), in the quote above, suggested that palaeontologists were not practising what Simpson (1944) and Mayr (1942), at least, had preached. Eldredge & Gould (1972) realised that the mechanism of speciation by geographic isolation demanded a new approach to the problem of evolutionary trends (Fig. 2.2), and a reorientation of palaeontology towards quantitative investigations of evolutionary patterns across whole faunas (Gould & Eldredge 1977).

One aspect of the argument of Gould & Eldredge's (1972) paper was almost immediately developed by Stanley (1975). He pointed out that evolutionary change would be concentrated in speciation events, and that the process of speciation is to a large extent random. Isolation, and hence peripatric speciation, arises accidentally, and often through the action of external agents. Additionally, populations are spatially heterogeneous, so the 'gene-pool' that is isolated is effectively randomly determined,

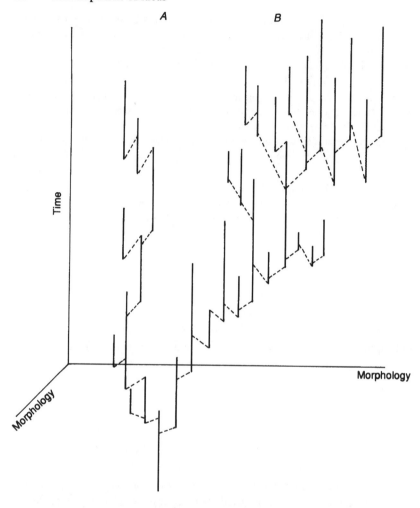

Figure 2.2 Three-dimensional sketch contrasting a pattern of relative stability (*A*) with a trend (*B*), where speciation (dashed lines) is occurring in both major lineages. Morphological change is depicted here along the horizontal axes, while the vertical axis is time. Though a retrospective pattern of directional selection might be fitted as a straight line in *B*, the actual pattern is stasis within species, and differential success of species exhibiting morphological change in a particular direction. Redrawn from Eldredge & Gould (1972, Fig. 5-10).

echoing comparison of speciation with mutation (Mayr 1963), which is an entirely random process. Thus,

if most evolutionary change occurs during speciation events and if speciation events are largely random, natural selection, long viewed as the process guiding evolutionary change, cannot play a significant role in determining the overall course of evolution. Macroevolution is decoupled from microevolution, and we must envision the process governing its course as being analogous to natural selection but operating at a higher level of biological organization *(Stanley 1975, p. 648)*.

Stanley (1975) used the term 'species selection' for this process operating on species and determining trends, as had de Vries (1906). He argued that 'natural selection seems to provide little more than raw material and fine adjustment of large-scale evolution', and the 'view that evolution can ultimately be understood in terms of genetics and molecular biology is clearly in error' (Stanley 1975), although this point is clearly not appreciated by all (see Cockburn 1991, quoted on pages 34–35).

Gould has continued to argue the case for punctuated equilibria passionately. He has described evolutionary time as a hierarchical system of distinct tiers: evolutionary events of the ecological moment (first tier), trends within lineages over geological time (millions of years) (second tier), and mass extinctions (third tier) (Gould 1985). In this terminology, the evolutionary processes that Darwin (1859) described operate at the first tier only. Gould has argued that:

our failure to find any clear vector of fitfully accumulating progress, despite expectations that processes regulating the first tier should yield such advance, represents our greatest dilemma for the study of pattern in life's history. I shall call it the *paradox of the first tier (Gould 1985, p. 4)*.

He presented two possible solutions: processes of the first tier, such as competition, do indeed regulate the history of life, but in ways that we do not currently understand; or else any progress that does accumulate is undone by processes of the second tier, which operate by quite different rules. Punctuated equilibria provides a partial solution to the paradox of the first tier because features are spread through lineages because populations that carry them have a greater tendency to speciate, and thus the spread is non-adaptive. There is then no reason to expect that evolutionary trends would be progressive. The first tier does not regulate events at the second, resolving the paradox.

Gould (1985) then advocated the idea that there is a distinctive third tier in the episodes of mass extinction that have generated faunal turnover at intervals of around 26 Myr throughout Earth history (Raup & Sepkoski 1984, 1986, 1988; Sepkoski 1989). There was, and still is, debate about the nature of mass extinctions and whether they are periodic and, if so, whether they have an extraterrestrial explanation (Davis *et al.* 1984; Rampino & Stothers 1984). It is not clear why these extinctions are experienced by fauna (especially marine invertebrates), but not by flora (Traverse 1988; Willis & Bennett 1995). The Earth is certainly subjected to all manner of extra-terrestrial phenomena, so, for example, the number and size of known asteroids in orbits that approach the Earth means that we can expect on average one collision every 1.4 Myr with objects greater than 2 km in diameter, one every 330 Myr with objects greater than 8 km in diameter, and so on (Shoemaker *et al.* 1990). However, the periodicity of mass extinctions has been questioned (Stigler & Wagner 1987, 1988; Heisler & Tremaine 1989; Patterson & Smith 1989; Quinn & Signor 1989), and a few, but not all, the alleged mass extinctions may be invalid statistically (Hubbard & Gilinsky 1992), so the matter is far from resolved (Glen 1994). Whatever the explanation, and whether or not there is a common, periodic, cause, these events are undoubtedly dramatic in impact and have far-reaching consequences for the course of life's history. They therefore may contribute to the resolution of the paradox of the first tier.

Gould's (1985) paper was significant for bringing together processes operating across a range of time-scales and arranging them in a unified hierarchical structure.

Species as individuals
There is currently philosophical debate about whether species are 'individuals' or 'classes' (Ghiselin 1974; Hull 1976, 1978, 1980). Logically, an 'individual' is not a synonym for 'organism': it means a particular thing, and can designate systems at a variety of levels. Individuals are localized in time and space, and may be made up of parts, which need not be identical, nor even contiguous: a human being is an individual, but so is Malaysia. It is characteristic of individuals that there cannot be examples of them, and they have proper names. Ghiselin (1974) compared species with businesses, such as IBM, which are also individuals. Although both have properties and can be described, their names cannot be defined. Correct usage demands 'Joe Bloggs is a specimen of *Homo sapiens*', not 'Joe Bloggs is a *Homo sapiens*', whereas the latter would be correct if '*Homo sapiens*' was a class. Higher taxa, such as 'mammals' are classes, however. Since species are made up of organisms related by descent, they need not even be composed of similar organisms, although they normally will be (Hull 1976).

Not everyone accepts the notion that species are individuals. But the philosophical notion that they are strikes a chord with the punctuated-equilibria interpretation of the fossil record. If Eldredge & Gould (1972) are right about that, then speciation is the all-important evolutionary process: species have 'births' and 'deaths' that are analogous with these processes for organisms. Above all, species are, in fact, spatiotemporally-bounded entities, which is exactly what the philosophical position states, and can even be sorted (Eldredge 1989). If species do not change during their existence, as Lyell (1832) and Eldredge & Gould (1972) argued (see pages 7 and 27), then species are not evolving. But they do form lineages, by speciation, and these can evolve (Hull 1980), creating the trends seen in the fossil record (Stanley 1975).

Turnover-pulse hypothesis
Elisabeth Vrba is a mammalian palaeontologist who has built on the model of punctuated equilibria (Eldredge & Gould 1972) by incorporating the effects of environmental change (Vrba 1985, 1992, 1993) into a 'turnover-pulse' hypothesis:

Evolution is normally conservative and speciation does not occur unless forced by changes in the physical environment. Similarly, forcing by the physical environment is required to produce extinctions and most migration events. Thus, most lineage turnover in the history of life has occurred in pulses, nearly synchronous across diverse groups or organisms, and in predictable synchrony with changes in the physical environment. Most of these turnover pulses are small peaks involving few lineages and/or restricted geographic areas. Some of them are massive and of global extent. The hypothesis invokes a combination of evolutionary notions: conservatism, habitat-specificity of species, vicariance and punctuated equilibria *(Vrba 1985, p. 232).*

The possible responses of species to environmental change are illustrated in Fig. 2.3. Vrba (1985) concluded that evolutionary change is largely a function of changes in the physical environment, especially climatic change, possibly of tectonic origin, but probably most often of astronomical origin. She noted:

the comparatively recent demonstration that changes in the earth's orbital geometry are the fundamental cause of the succession of Quaternary ice ages [(Hays, Imbrie & Shackleton 1976)], and a proposition that astronomical cycles have operated right through biotic evolution, place a new onus on evolutionists *(Vrba 1985, p. 234).*

32 · Development of ideas

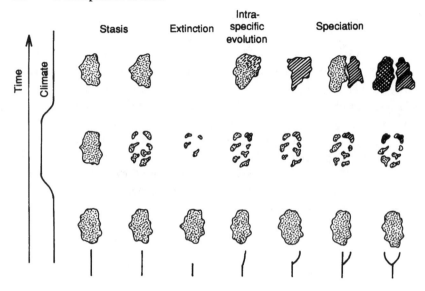

Figure 2.3 Alternative responses of species to climatic, or other environmental, change. Irregularly outlined shapes represent the distributions of species. Simple phylogenetic trees are placed beneath each response. Redrawn from Vrba (1985, Fig. 1).

She added that it is difficult to square the view of gradual evolution of species over millions of years with climatic oscillations over time-scales that are short compared to species durations.

In similar vein, Futuyma (1987) pointed out that shifting habitat locations, such as those forced by Quaternary climatic changes, mean that any differentiation of local populations is typically ephemeral. On the other hand, reproductive isolation confers sufficient permanence on morphological changes for them to be seen subsequently in the fossil record. These observations nicely tie peripatric speciation ideas in with the notion of environments that are ephemeral relative to species duration, and which are known to occur at least during the Quaternary.

Vrba (1993) described observations on the late Cenozoic and modern record of African bovids, leading to the conclusions:

1. Single species commonly occur across different ecosystems at any one time. Populations of a species share the same habitat-specificity but not necessarily the same biotic context.
2. The community–associations of particular species have changed through time with climatic cycles over time scales that are short with respect to most known species durations [Fig. 2.4].

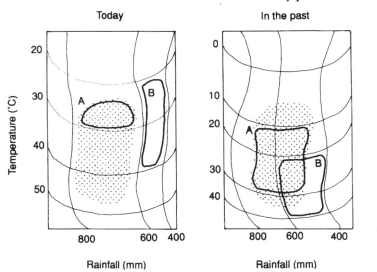

Figure 2.4 Hypothetical illustration of how two species, A and B, with overlapping tolerances for three habitat variables can have disjunct distributions today, but sympatric distributions in the past. Two substrates are shown as stipple and blank, and two climatic variables vary across the region, as shown. Species A occurs only on the stipple substrate, with temperature tolerance of 10–30°C, and rainfall tolerance of 400–800 mm. Species B occurs on either substrate, with temperature tolerance of 20–40°C, and rainfall tolerance of 400–600 mm. Redrawn from Vrba (1993, Fig. 3).

3. Subparts of the same species have different ecological histories, because biotic interactions in different ecosystems that share one or more species are likely to run different courses .... Such different ecological and evolutionary regimes, between conspecific populations, can continue in the presence of substantial gene flow between those populations ....

4. Most changes in species composition in ecosystems do not correspond to changes in genealogical species diversity. (Yet each change in the latter must be reflected by a change in the former, at least in one ecosystem.) The extensive evidence of large-scale drift of species' geographic distributions suggests that a large proportion of first and last appearances of taxa in particular areas has nothing to do with speciation and extinction.

Whatever we may deduce about evolution in one ecosystem does not necessarily extrapolate to long-term evolution, namely to speciation, extinction, and to trends in phylogenies *(Vrba 1993, pp. 426–427).*

The 'turnover-pulse' hypothesis represents a strong form of environmental control of evolution, also emphasized by Haffer (1990).

'Modern synthesis' versus 'punctuated equilibria'

Much of the comment on punctuated equilibria has been centred around demonstrating that neither Darwin nor any of the modern synthesis authors ever said that they thought that evolution took place by continuous gradual change over long periods of time. Dawkins (1986) and Levinton (1988) devoted several pages to showing that rates of evolution vary over time, can be sufficiently rapid as to appear instantaneous in the fossil record, and that no-one ever denied that this was the case. They are certainly right to point this out, but the interesting point about the punctuated equilibria hypothesis is to do with questioning whether modern processes are all that is necessary to explain the evolutionary record, and that is the key area where the differences between the the defenders of the modern synthesis and the architects of punctuated equilibria become stark. Eldredge & Gould (1972) have been accused of setting up a straw man in their definition of phyletic gradualism (Grant 1985; Levinton 1988), but defenders of the modern synthesis are guilty of the same in focusing their criticism of punctuated equilibria on rates of evolution rather than on its challenge to their uniformitarianism. The core of the debate is thus over the question of whether knowledge of the present is adequate to explain the record of the past, and the only way of testing this position is in the fossil record.

Use of the fossil record as a source of evidence on this question is rare. Influential books on evolution over the last 20 years have either ignored the fossil record completely, or used it simply as a source of evolutionary 'Just So Stories' to exemplify points being made from observations of modern organisms and ecosystems (for example Maynard Smith 1975; Dawkins 1976, 1983, 1986, 1995). Recent textbooks contain statements such as:

The synthetic theory can be characterized as the population genetical approach to microevolution and its extensions to other evolutionary levels and to other biological fields. In its core it represents a combination of the population geneticist's approach, which provides theoretical precision, with the naturalist's approach to living populations and species, which brings in the touch with reality. In its entirety it encompasses a much larger range of fields. Thus considered, it is not a special theory, which can be verified or falsified, but a general theory, a paradigm, which can absorb changes and modifications within wide limits, and has done so over the years since its inception *(Grant 1985, pp. 16–17).*

The final criticism which can be directed at the punctuated equilibria model is its emphasis on the founder event model of speciation. Because Mayr's original model attracts little favour from geneticists, it is a shaky basis on which to build a general theory of evolutionary change *(Cockburn 1991, pp. 248–249)*.

The best understood macroevolutionary transitions suggest that macroevolution occurs by extrapolated microevolution; the evidence does not suggest that any kinds of developmental change characteristically happen any more frequently in the origin of higher groups than in smaller-scale evolution *(Ridley 1993, p. 551)*.

Attitudes such as these, including the substantive uniformitarianism, shrugging off much of recent debate in palaeontology, not to mention much of the fossil record, indicate that in many quarters there has been little change in the last 50 years. Charles Lyell, a geologist, queried how it is that species remain substantially unchanged in the face of major climatic changes if natural selection is continually adapting populations to their environments. His observation has been borne out by considerable further work in the last 150 years, and the question he put to Darwin on this subject (see page 9) remains valid. It appears that no evolutionary ecologist has ever addressed it: most appear either ignorant of the problem, or ignore it, preferring to stick with a rigid modern-synthesis approach. Lyell was instrumental in establishing the view that Earth history could be explained in terms of processes now operating, and Darwin was surely influenced by Lyell's arguments. The Darwinian and modern synthesis views of evolution are characterized by explaining the fact of evolution in terms of present processes (substantive uniformitarianism). But Lyell was a close observer of the fossil record, and used it to help him distinguish which modern processes were influential in determining the past record. This remains a principal reason for examining the fossil record, warts and all.

Ecological processes

How does ecology, as practised today, compare with the development of ideas in evolution? What, in fact, is ecology? The *Shorter Oxford English Dictionary* defines it as 'that branch of biology which deals with the relations of living organisms to their surroundings, their habits and modes of life, etc.'. Most textbooks give a similar definition, and at this level there is no disagreement about what ecologists do on a day-to-day basis. However, the original definition by Haeckel, in 1866, added that 'ecology is the study of all those complex interrelations referred to by Darwin as the conditions of the struggle for existence' (Allee *et al.* 1949, frontispiece). *On the Origin of Species* (Darwin 1859) predates Haeckel's definition, but can still be considered as the first ecological textbook, because it did set the research paradigm that has, implicitly, dominated ecology ever since: evolution by means of natural selection. Darwin was, after all, insistent that competition was the dominant factor in the natural selection process (see Darwin 1859, Chapter 3).

The scope of ecology was outlined by Darwin (1859), but it did not expand into a scientific discipline until the late nineteenth century, contemporaneously with genetics (McIntosh 1985). Rather than developing as a unified discipline, ecology then split promptly into subdisciplines (animal ecology, plant ecology, limnology, and oceanography), which have generally proceeded independently ever since. Other subdisciplines, notably mathematical ecology, have arisen more recently. Ecology has never functioned as a whole, and, numerous textbooks notwithstanding, has never had a synthetic work (other than *On the Origin of Species*), and there have been no advances or unifying developments affecting all subdisciplines. There are, indeed, deep problems about the way ecologists go about their business, both on an academic level (Peters 1991), and by confusion with political and religious movements.

One work that came nearest, single-handedly, to advancing and defining ecology was the text *Animal Ecology* by Elton (1927). He began by defining ecology disparagingly as 'scientific natural history', but then developed principles of food chains and the food cycle, food size, the niche, and the pyramid of numbers (McIntosh 1985), before finishing with a chapter on evolution, although 'it may at first sight seem out of place to devote one chapter of a book on ecology to the subject of evolution' (Elton 1927, p. 179), perhaps indicating that the two disciplines were already completely distinct. Elton was interested especially in animal numbers and changes in population sizes, and used this interest to develop a distinctive angle on ecology and evolution. He described the argument in favour of natural selection, but then cited evidence that suggested that colour dimorphisms found in mammal species were not adaptive, and that the characters separating closely related species also did not appear to be adaptive. There must, therefore, be some process that allowed the spread and maintenance of non-adaptive characters in populations (Elton 1927). He then pointed out that the theory of natural selection:

says that all animals tend to increase, and at a very high rate, but are prevented from doing so by checks. What has been said about fluctuations in numbers shows that such is not always the case. Many animals periodically undergo rapid increase with practically no checks at all. In fact, the struggle for existence sometimes tends to disappear almost entirely. During the expansion in numbers from a minimum, almost every animal survives, or at any rate a very high proportion of them do so, and an immeasurably larger number survives than when the population remains constant. If therefore a heritable variation were to occur in the small nucleus of animals left at a minimum of numbers, it would spread very quickly and automatically, so that a very large proportion of numbers of individuals would possess it when the species had regained its normal numbers. In this way it would be possible for non-adaptive (indifferent) characters to spread in the population, and we should have a partial explanation of the puzzling facts about closely allied species, and of the existence of so many apparently non-adaptive characters in animals *(Elton 1927, pp. 186–187)*.

This view should be contrasted with the later opinion of Mayr (1942) on the overall balance in population numbers (see page 20). Elton showed how an ecological approach could yield a distinctive angle on the process of microevolutionary change. His work preceded the modern synthesis, of course, but its implications are clear. Moreover, population ecology was not one of the disciplines drawn into the synthesis by Dobzhansky, Mayr, or Simpson.

A quite separate development came in concepts of communities developed among American plant ecologists working during the first part of the twentieth century. Early work saw vegetation formations and associations as climatically-determined climaxes, or as being in the course of successional development towards such climaxes (Weaver & Clements 1929). An alternative approach, radical and derided at the time, was the 'individualistic' concept of species within plant associations:

it may be said that every species of plant is a law unto itself, the distribution of which in space depends upon its individual peculiarities of migration and environmental requirements *(Gleason 1926, p. 26)*.

Gleason's view was ignored for 20 years, but was eventually rehabilitated by a newer generation of plant ecologists, such as Egler (1947). It was developed from study of modern vegetation and communities, but foreshadowed results from pollen-analytical investigation of the dynamic behaviour of plant communities during the last 10 kyr (see Chapter 5).

The view of evolution espoused by most ecologists today is basically that of Darwin (1859) overlain by modern understanding of genetics. For example, Bradshaw (1984) gave three major reasons for considering evolution as a proper subject for ecologists:

(i) We all tacitly assume that what we examine ecologically is the product of evolution, and that as a result of natural selection it is adapted, more or less, to its environment, in the sense that it is fitted (*aptare*) to (*ad*) it.

(ii) Species are clearly not fixed, but consist of a complex of different populations, often with extremely different ecological properties, which can change as a result of evolutionary processes in only a few generations.

(iii) Ecologists themselves study life and death, and mechanisms of fitness, which are the stuff of evolution as we understand it *(Bradshaw 1984, p. 1).*

Pianka (1988, p. 5) asserted that 'the single concept closest to deserving the status of "law" in ecology and one that is shared with all of biology, is **natural selection**' (emphasis in original). Harper (1967) made a ringing and explicit plea for plant ecology to return to Darwinian principles, and Hutchinson (1959) was a self-proclaimed believer in natural selection as the process contributing to evolutionary progress. Much of the practice of ecology and evolution today is concerned directly with adaptation or natural selection in some way, and the two disciplines have always been closely associated in the minds of ecologists (Collins 1986). However, the fossil record is rarely mentioned, let alone used, although it is implicit that evolutionary ecology contributes to the second part of Darwin's great argument. Demonstration and documentation of the fact of evolution has been left with geologists, while the process of evolution is studied by ecologists in woods, ponds and bug bottles. The processes of change are assumed to be completely understandable from the functioning of modern systems (Eldredge 1992). It is important for much work in evolutionary ecology to discover whether work on current systems does or does not contribute to understanding of the evolution of life and the degree to which its extrapolation from ecological to geological time-scales is justified. Since substantive uniformitarianism has been generally rejected in other disciplines (Gould 1965), the onus should now be on evolutionary ecologists to justify their approach, rather than for others to show that it is invalid. Does evolutionary ecology, perhaps, have more to do with the maintenance of life at or near its current state? There is a great temporal void here, and it can only be filled by data collected at longer time-scales than most evolutionary ecologists are currently examining.

MacArthur (1972) considered that unravelling the history of a phenomenon and describing the machinery of a phenomenon were two distinct processes, and that individual natural scientists tend to work with one or the other, but seldom excel at both. He suggested that ecologists (and physical scientists) are machinery-orientated, whereas biogeographers and palaeontologists are history-orientated, and the two groups notice different aspects of the natural world (MacArthur 1972). MacArthur had earlier made a similar distinction in a discussion of patterns of species diversity, arguing that:

if the patterns were wholly fortuitous and due to accidents of history, their explanation would be a challenge to geologists but not to ecologists. The very regularity of some of the patterns for large taxonomic groups suggests, however, that they have been laid down according to some fairly simple principles *(MacArthur 1965, p. 510).*

The conjunction of 'fortuitous' and 'history' suggests that MacArthur thought that patterns with an historical explanation were not susceptible to scientific study. *On the Origin of Species* may have been a prototype for an ecology textbook, but remains unusual for blending both the historical and mechanistic aspects of the subject. If most ecologists are indeed machinery-orientated, then the reason for a return-to-basics plea (Harper 1967), and a restatement of principles (Bradshaw 1984: and see page 38) becomes clearer.

Quaternary research

Understanding of the events of the Quaternary period has been developed by geologists and ecologists into a distinct field of study with its own traditions and practices. The source material is geological, but often unconsolidated, and the fossils are of modern species. As a result, most research into aspects of this period are published in 'Quaternary' journals, and not the mainstream scientific literature of either geology or biology. Biologists working on Quaternary material typically refer to themselves as 'palaeoecologists' rather than palaeontologists. No such distinctions existed in the early days of research into the recent past: Charles Lyell's first geological paper was on Holocene lake sediments (Lyell 1826). How has our knowledge of the period arisen, and how has research into the Quaternary become separated from both geology and ecology?

Nineteenth century geologists were fully aware that the Earth, and especially the northern hemisphere, had been subjected to subcontinental-scale glaciation in the not too distant past, following the researches of Agassiz (1840) in the Alps and elsewhere. At the time, the idea was radical, and Agassiz concluded his book with a plea:

Quelque opposition que l'on puisse faire aux idées énoncées dans cet ouvrage, toujours est-il que les faits nouveaux et nombreux que j'y ai consignés, surtout relativement à l'état intérieur des glaciers, à leur action sur le sol et au transport des blocs erratiques, ont amené la question sur un autre terrain que celui sur lequel elle a été débattue jusqu'à présent *(Agassiz 1840, p. 330)*.

Whatever the opposition against the ideas discussed in this work, it is unquestionable that the numerous and new facts that I have presented, particularly with respect to the internal conditions of glaciers, to their action on the substratum, and to the transportation of erratic boulders, have completely changed the context in which the question has been discussed up to the present *(translation from Carozzi (1967, p. 177))*.

The idea was not universally accepted for another 20–30 years, and where it was (Darwin 1859), the language used was typically in terms of 'the Glacial Period', treating it as a single block of time, although it was certainly appreciated that there had been a number of glacial fluctuations. Geikie (1874) detailed evidence for an interglacial period, with a climate at least as warm as present, within the 'Glacial epoch', adding 'we cannot yet say how often such alternations of cold and warm periods were repeated'. His table showing the stratigraphic record of Britain includes three stages within the 'Glacial epoch': the last glacial period, the last interglacial period, and, earliest, a 'Great cycle of glacial and interglacial periods'. It was to be another century before deep-sea sediments revealed the full-extent of glacial–interglacial oscillations and provided a time-scale (Shackleton & Opdyke 1973), discussed in Chapter 3.

Darwin and his contemporaries were, therefore, fully aware of the probable existence of former glaciation, and that some species had survived throughout without modification. Both Lyell (see page 9) and Darwin (see page 13) commented on this, in different ways. But they were completely unaware of the frequency of climatic oscillations of the Quaternary in absolute terms, or relative to the durations of species in the geological record. In the first edition of *On the Origin of Species*, Darwin (1859) speculated on the age of the Earth, reaching a figure of about 300 million years for the denudation of the Weald in southern England, and hence a period of time much greater than this for the appearance and development of the forms present before the Wealden sediments were deposited. Thomson (1862) provided the first scientific estimate of the age of the Earth by analysing the rate of cooling of the Earth since its formation. He obtained a figure of between 20 and 400 million years. This figure was considered highly authoritative, and stood for three decades as the best physics could offer (Dalrymple 1991). This did not seem like enough time to Darwin, and he speculated that evolution might have proceeded faster when the Earth was subjected to more rapid and violent changes in its physical conditions (Darwin 1872). The best contemporary data on the timing of the 'Glacial period' came from Croll's calculations of orbital variations, and these placed it between 80,000 and 240,000 years ago (Croll 1868). Development of a more accurate time-scale for the age of the Earth and its geological succession followed on from the discovery of radioactivity. The application of radiometric age determinations during the early twentieth century showed, by the 1920s, that the age of the Earth was billions of years, not tens of millions as Thomson (1862) and others had argued (Dalrymple 1991). The present accepted age is 4.54 Ga, supported by radiometric and other lines of evidence (Dalrymple 1991), and a detailed geological time-scale is available (Harland et al. 1990: and see Table 1.1). Within the Quaternary, a range of dating methods is now available (Smart & Frances 1991), of which much the most useful is the radiocarbon method (see Chapter 5). This technique provides age determination on organic sediments more recent than about 40 ka. It thus covers the transition from the last glacial stage into and during the present interglacial. Age determinations by this method have been available for the last 30–40 years, but it is only within the last 20 years that an accumulation of determinations have enabled the process of examining events during an interglacial to be carried out using a consistent and generally applied time-scale rather than relative litho- or bio-stratigraphy. This key technical advance has meant that part of the geological column now has a rigorous and accurate time-scale with a potential resolution of hundreds of years. The quantity of information that has resulted, however, has been accompanied by an expansion of terminology and technique not used in other parts of the geological record. Rather than integrating Quaternary research fully within geology as the one part of the record with such fine temporal resolution, the result has been the separation of the period into a discipline of its own: too close to the present for geologists, and too geological for ecologists.

Much of the effort of Quaternary palaeoecology within the last few decades has been spent on reconstructing past climates, especially those of the Holocene, and very little on the ecological and evolutionary consequences of observed patterns of behaviour of plants and animals on Quaternary time-scales, although there are significant exceptions (for example Coope 1979). It had been recognized since the early twentieth century that changing abundances of pollen in suitable sediment sequences, and hence forest composition, may be related to changing climates, and the first Holocene pollen stratigraphic schemes were based around a presumed synchroneity of vegetation change and climatic change (for example Godwin 1934; von Post 1946). But this elegantly simple explanation subsequently fell into disfavour as it became clear that other factors, including at least pathogens (Watts 1961; Davis 1981a), soils (Brubaker 1975; Pennington 1986), and human activities (Iversen 1941), also played a role in controlling the relative abundances of the pollen-producers in Holocene forests. Attempts at mapping changing abundances of pollen types using the radiocarbon time-scale (for example Davis 1976; Huntley & Birks 1983) were influential in demonstrating the broad-scale individualistic nature of species behaviour on Holocene time-scales. They indicated that the spread of trees during the early Holocene (and hence appearance and increase at any one site) might have much more to do with distance from glacial refugia and rates of spread than with close tracking of climatic change (Birks 1981). By this time the possibility of using pollen data as a proxy for past climate change was receiving considerable attention because of the increasing potential of numerical methods (Webb & Bryson 1972). The interest in using pollen data in this way arose because of the availability of continuous records on a continental scale. However, there was never a demonstration that vegetation (or pollen) did track climatic change closely enough for this to be a reasonable approach. Indeed, making such a demonstration would require independent data on past climate, and if that was available there would be no need to use pollen as a proxy.

Debates about the use of pollen-analytical data for reconstructing past climates absorbed much of the attention of Quaternary palaeoecologists through the last 20 years, whether arguing for or against the notion. As a result, the significance of data collected during that period for understanding of evolution and ecology on time-scales of thousands of years was largely not appreciated, either within the Quaternary scientific community or beyond it.

Discussion

Although an accurate geological time-scale was not available to Lyell or Darwin, subsequent generations have known the overall time-scale for the evolution of life. But no period of Earth history had an accurate enough chronology at time-scales fine enough to resolve distribution changes within the duration of species until the advent of generally-available radiocarbon age determinations, and that did not happen until well after the development of the modern synthesis, and just after the punctuated equilibria debate was initiated by Eldredge & Gould (1972).

Evolutionary theorists have thus produced two contrasting hypotheses about the response of species to Quaternary climatic changes. Darwin (1859, 1872) argued that species would remain unchanged as long as they responded together, but change could be expected if species were unable to maintain the same mutual relations. Mayr (1942) argued that speciation would be promoted by climatic change associated with glaciation. On the other hand, Lyell (see page 9) and de Vries (1906) suggested that species survive such climatic change unmodified, despite experiencing altered physical and, or, biotic environments. Both cannot be right: does evolution normally take place in response to these climatic changes, or does it not? Or is the answer 'sometimes'? Supporters of phyletic gradualism and punctuated equilibria have argued over rates of evolution. But the real test is what happens when environments do change. Under phyletic gradualism, the expectation is that there would be evolutionary change, but under punctuated equilibria there is not necessarily an expectation of change. What does the fossil record show during periods (such as the Quaternary) of undoubted environmental change?

Ideas about how life evolved have developed erratically through the course of the twentieth century. This development has been paralleled by an expansion of the understanding of events of the last million of years or so, providing increased temporal resolution of the behaviour of species over time-scales much less than millions of years, and down to hundreds of years in favourable circumstances. This knowledge of Quaternary events and time-scales impinges on both evolutionary theory and ecological theory, but has not yet been incorporated into either, even though the importance of events within the last period of glaciation was recognized by Lyell, Darwin, and de Vries, among others (see above). Why has this incorporation not taken place? Is it coincidence that this subject was developing at a time when evolutionary theorists were engaged in their own internal debates over phyletic gradualism and punctuated equilibria?

3 · Orbital-forcing of climatic oscillations

All processes, geological and biological, taking place at the surface of the Earth are controlled and influenced by local, regional, and global climate. In this chapter, the way in which climates change with time is described, together with the reasons for thinking that the patterns observed for the last few million years are likely to be a reasonable representation of the way in which climates have changed throughout Earth history.

Solar insolation is the main source of heat for the surface of the Earth and a crucial element in global climate systems today and in the past. Diurnal and annual periodicities in temperature and other climatic parameters are familiar consequences of life on a planet that spins about an axis that is inclined to the plane of its orbit around its heat source. The Earth's location and velocity in space are critical for determining the amount of solar insolation received at its surface, and the latitudinal distribution of that insolation. However, the Earth's orbital characteristics arise from gravitational attractions, dominated by the Sun, but also including gravitational attractions between the Earth, the Moon, and the other planets. The Solar System is a dynamic system: the location and velocity of any of its elements at any point in time is influenced by the locations and velocities of all the others, to at least some extent. This system is, in general, deterministic, but it is currently not possible to obtain a general relationship that enables the determination of the location and velocity of all the planets for all times, past, present and future. Approximations need to be made for all cases of this type, involving three or more elements. This chapter considers the nature of changing orbital parameters, their consequences for insolation received by the Earth, and the modelling of climatic changes from changing insolation values (Berger 1988, 1989). A general understanding of this system is essential to a full appreciation of the nature of life on a planet that has, for billions of years, been spinning and weaving its way around the Sun, completely under the influence of the gravitational pulls of other moving bodies on its own mass, in a complex 'dance' (Imbrie & Imbrie 1979).

Although the topics in each of the sections of this chapter are treated distinctly, all have developed, especially over the last two decades, as a result of close collaboration between astronomers, geophysicists, climatologists, and geologists. The interplay between these groups has been crucial in developing today's understanding of orbital variations and their consequences.

Orbital parameters

The Earth today has an axial tilt of about 23° 27', and an elliptical orbit with an eccentricity, e, of 0.016 ($e = (a^2 - b^2)^{1/2}/a$, where a and b are the semi-major and semi-minor axes of the ellipse, respectively). Perihelion (the point on the Earth's orbit when it is nearest to the Sun) is 3 January, close to the date of winter solstice in the northern hemisphere (Imbrie & Imbrie 1979; Berger 1988). It has been known for 2000 years that variations occur in the axis of the Earth's rotation, and during the last 200 years it has been realized that variations also occur in eccentricity, obliquity (or axial tilt), and precession (the movement of perihelion relative to the vernal equinox). Calculations of the magnitude of changes have been made (Berger 1978a, 1984, 1988; Imbrie & Imbrie 1979), by solving analytically a set of differential equations describing planetary motions. The most accurate solution to date is that of Berger (for example 1978a) replacing an earlier solution by Vernekar (1972). The main orbital parameters (eccentricity, precession, and obliquity) are expressed as quasi-periodic functions of time, allowing computation of the main periods of orbital variation, which are 400, 100, 41, 23, and 19 kyr (Fig. 3.1). These are mostly due to the influence of Venus (because it is close to Earth) and Jupiter (because of its great mass), and to the orbital precession (Berger 1988). During the last 5 Myr, eccentricity has varied from 0.0 (circular orbit) to 0.0607, with a primary period of about 100 kyr and a secondary period of about 400 kyr; obliquity has varied between 22° and 24° 30', with an average period of about 41 kyr; and the precession element has a major period, relative to the moving perihelion, of about 23 kyr and a minor period of about 19 kyr (Berger 1988). These are, of course, only the dominant periods of orbital variation: a fuller expansion can be found in Berger (1978b).

Because of the complexity of Solar System dynamics and because of possible unknown complications on long time-scales, it is not yet possible to extend these calculations back in time for more than about 5 Myr (Berger 1984; Berger & Pestiaux 1984). Using another approach, Berger

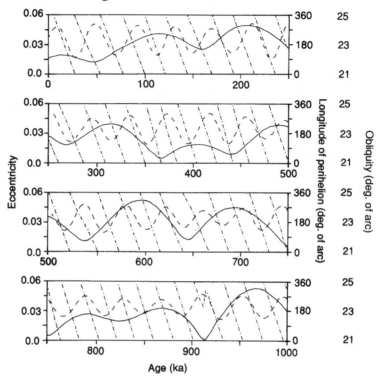

Figure 3.1 Variation of the eccentricity (continuous line), obliquity (dashed line), and longitude of perihelion (precession of the equinoxes: dotted and dashed line) of the Earth's orbit since 1 Ma. Redrawn from Berger (1976, Fig. 2b).

et al. (1989a, b) considered long-term variation in the Earth's orbital elements by investigating the interplay between various elements and their dependence on three parameters (Earth's rotation speed, Earth–Moon distance, and the Earth's shape) that influence the orbital elements and for which there are geological data on variation over time. They found that the periods of precession and obliquity have increased over the past 2.5 Gyr to their modern values (Table 3.1), but remain within the same order of magnitude.

Laskar (1989, 1990), however, used a modified numerical integration technique to explore the behaviour of the Solar System over 200 Myr. The orbital motion of the inner planets is too rapid for direct numerical integration to be reliable, but he did find that the inner Solar System, including the Earth, exhibits quasi-periodic behaviour on time-scales of

Table 3.1. *Estimated values for periodicities (kyr) of orbital parameters, taking into account the effect of variation in the Earth–Moon distance and variation in the length of day*

Date (Ga)	Precession		Obliquity	
0	19.000	23.000	41.000	54.000
0.5	19.117	23.171	41.550	54.958
1.0	19.075	23.111	41.353	54.614
1.5	18.988	22.983	40.945	53.905
2.0	18.832	22.754	40.224	52.662
2.5	18.573	22.377	39.060	50.684

Source: From Berger *et al.* (1989a, Table 6).

10 Myr. At longer time-scales, predictability breaks down, and the system exhibits chaotic behaviour (Laskar 1989). A more recent numerical integration of the entire Solar System for 100 Myr has confirmed that the evolution of the system as a whole is chaotic (Sussman & Wisdom 1992).

Geological evidence (see Chapter 4) indicates that there have been pre-Quaternary climatic oscillations with periods of the same order of magnitude as today. Thus, even though the behaviour of the system is chaotic for predictions on time-scales longer than tens of millions of years, it is clear that pre-Quaternary time has been characterized by oscillations of similar periodicity to the Quaternary. It may yet be possible to make predictions about behaviour at shorter time-scales of Earth history, at least for the Phanerozoic. However, necessary evidence for the Earth's rotation speed, distance from the Moon, as well as the Earth's shape and the state of its interior, are not yet available for earlier eons (Berger *et al.* 1989a), as values for these parameters in remote periods are not necessarily the same as today.

Changing orbital variations on time-scales of 20–400 kyr are thus a fundamental consequence of the existence of the Earth in a universe controlled by gravitational forces, and of a Solar System consisting of several planets with their particular sizes and configurations: a contingent fact of Earth history. The other planets exhibit similar behaviour, for the same reason, but have been less studied. Changing orbital parameters for Mars, for example, are shown in Fig. 3.2.

48 · Orbital-forcing of climatic oscillations

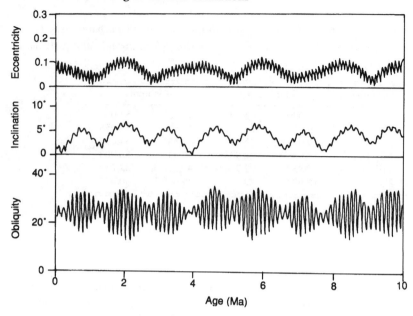

Figure 3.2 Orbital variations of Mars. Redrawn from Ward (1979, Fig. 1).

Insolation

The extension of the theory of orbital variations to the calculation of insolation is due to Milutin Milankovitch, a Serbian engineer, who dedicated 30 years to the task in the first half of the twentieth century. In so doing, he generated considerable interest in the question of astronomical forcing of the Earth's climate, and produced testable hypotheses about the geological record of past climate (Imbrie & Imbrie 1979). Orbital variations in the 10–400 kyr band are now often termed 'Milankovitch cycles' (oscillations), as a tribute to the extent of his contribution. The Earth's orbital variations affect seasonal and latitudinal solar radiation receipts at the top of the atmosphere. When eccentricity is zero, there is no difference in solar radiation receipt between the two halves of the year. At maximum eccentricity, receipts vary by about 30%, affecting the relative intensity of the seasons, and in an opposite sense in the two hemispheres (Bradley 1985). Changes in obliquity define the latitude of the polar circles and tropics, with greatest effect at high latitudes, but with equal effects in both hemispheres. Precession affects the timing of the seasons relative to perihelion and aphelion, and is thus modulated by eccentricity (Imbrie & Imbrie 1979; Bradley 1985). If summer occurs

Insolation · 49

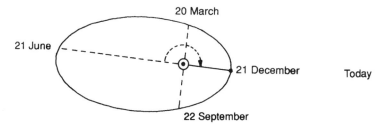

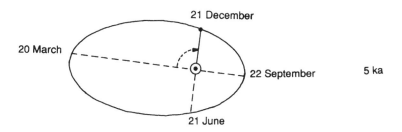

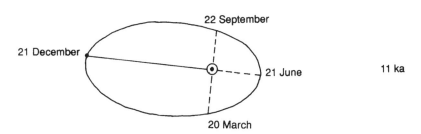

Figure 3.3 Precession of the equinoxes. Owing to axial precession and to other astronomical movements, the positions of equinox (20 March and 22 September) and solstice (21 June and 21 December) shift around the Earth's elliptical orbit, completing one full revolution about every 22 kyr. At 11 ka, the winter solstice occurred near one extreme of the orbit. Today, the winter solstice occurs near the other extreme. As a result, the Earth–Sun distance on a given date changes. •, Earth's position on 21 December; ⊙, Sun. Redrawn from Imbrie & Imbrie (1979, Fig. 16).

while at perihelion (and hence winter at aphelion), seasonal contrasts are enhanced relative to the converse situation (Fig. 3.3). The timing of perihelion is irrelevant when eccentricity is near zero, but an important feature with high eccentricity (Bradley 1985).

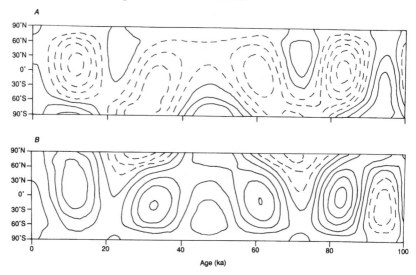

Figure 3.4 Long-term variations of deviations of received solar radiation from their modern values, for the caloric northern hemisphere winter and southern hemisphere summer (*A*), and for caloric northern hemisphere summer and summer hemisphere winter (*B*). Contour interval is 10 cal cm^{-2} day^{-1}. The continuous lines are for positive deviations and dashed lines for insolation below modern values. Redrawn from Berger (1978a, Fig. 2a).

Modern calculation of the distribution of insolation by latitude and season is due to Berger (for example 1978a: and see Fig. 3.4). The dominant forcing factor for insolation at high latitudes is clearly the obliquity term, while at low latitudes, eccentricity and precession terms dominate. Variation occurs across the full range of latitudes, from the tropics to the poles. It is seen to be unique, in terms of latitudinal variation and rate of change, at any moment in time. The diagrams in Fig. 3.4 reflect changing insolation received by the Earth at the top of the atmosphere, and are therefore not affected by any set of circumstances at the surface of the Earth itself, such as continental configuration, mountain-building, or the existence of continental ice-sheets. Since the Earth has always been subject to orbital variations, there must always have been changes in the pattern of insolation receipts in the manner displayed in Fig. 3.4, although differing in detail.

Climate models

How does variation in insolation pattern translate into changing climates at the Earth's surface? Access to information on past climates depends on interpretation of geological records. This was once done qualitatively, but interpretations are now made by various quantitative means. Once a record has been used as the basis for climatic reconstruction, however, it is no longer possible to compare that record with the inferred climate; to do so would be a circular argument. Where the interest is in the responses of organisms, communities, or environments to past climates, it is essential that the climatic record has been obtained independently from the geological record. Attempts are now being made to develop computer models of the global climate system that can be used to make predictions of likely past or future climates under conditions different from today. Early computer experiments along these lines did not include the effects of climatic forcing by orbital variations, but the current generation of General Circulation Models (GCM) do (Street-Perrott 1991). These models fall into two categories: those that attempt to produce a 'realistic' view of the Earth's climate at a particular time (or times) in the past, and those that explore the sensitivity of models by changing one or more elements (Street-Perrott 1991). The most detailed 'realistic' model is that for the past 18 kyr presented by Kutzbach & Guetter (1986), refined slightly by Kutzbach & Gallimore (1988), and subsequently revised (Kutzbach et al. 1993), using the Community Climate Model (CCM) at the National Center for Atmospheric Research, USA (NCAR). All such models compute certain climatic variables given certain sets of conditions, usually called boundary conditions. In the case of the CCM, boundary conditions included modern topography, modified as necessary for the distribution of ice-sheets and sea-levels (based on geological data), sea-surface temperatures and distribution of sea-ice (from marine core data), and albedo (surface reflectivity). Albedo was specified for different types of ground cover. Insolation was calculated from the orbital parameters obtained by Berger (1978a). Carbon dioxide levels were taken to be 330 p.p.m. throughout, with an additional experiment run using 200 p.p.m. for 18 ka, closer to full-glacial conditions. The model was run for a grid across the globe of about 4.4° latitude by 7.5° longitude, at 3-kyr intervals.

Selected maps (see Fig. 3.5) display the way in which global climates have changed in relation to the orbital variations of the last 18 kyr (see also Kutzbach & Guetter 1986; COHMAP members 1988). There are statistically-significant departures from present values of temperature and

52 · Orbital-forcing of climatic oscillations

A

July temperature (°C)

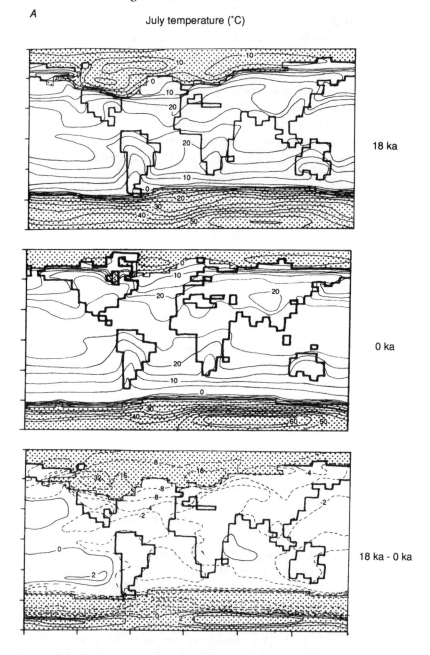

Figure 3.5 A–D. Simulated 18 ka and present climatic variables, and differences between them from GCM modelling experiments. Contours on the difference diagrams are dashed for negative values (present values greater than 18 ka values). Stippled area indicates extent of continental ice-sheets and floating ice. Continental landmasses are outlined in heavy, continuous lines. Redrawn from Kutzbach *et al.* (1993, Figs 4.1–4.4 and 4.9–4.12).

Climate models · 53

B
July precipitation (mm day⁻¹)

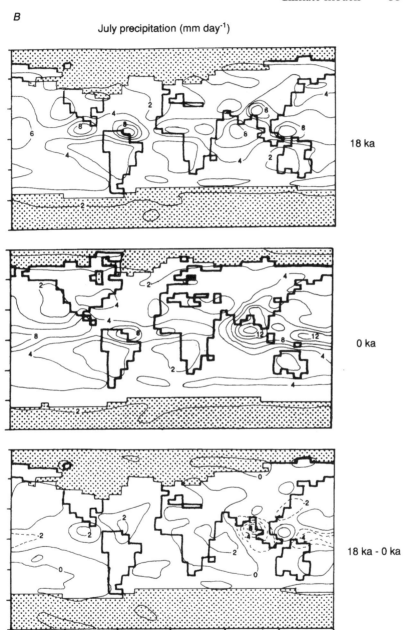

18 ka

0 ka

18 ka - 0 ka

Figure 3.5 (continued) For legend see page 52.

C January temperature (°C)

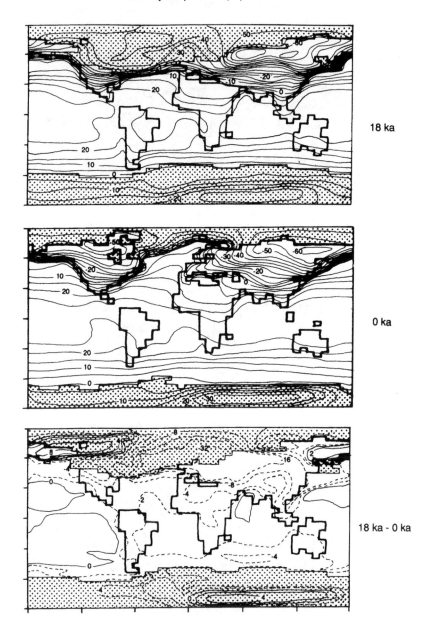

Figure 3.5 (continued) For legend see page 52.

Climate models · 55

D January precipitation (mm day⁻¹)

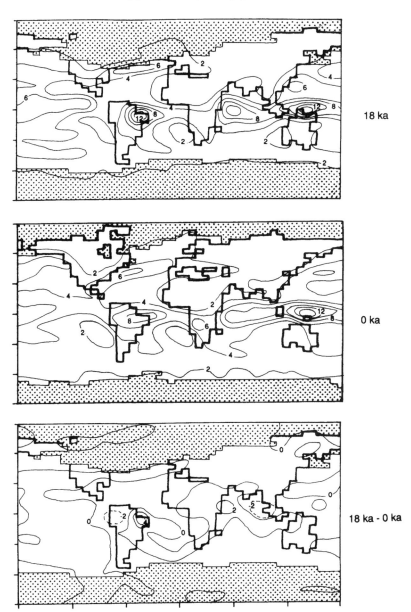

18 ka

0 ka

18 ka - 0 ka

Figure 3.5 (continued) For legend see page 52

precipitation, summer and winter, in all regions of the globe. Temperature changes were most extreme near the northern hemisphere ice-sheets, but there were also significant changes in tropical areas, remote from the direct influence of continental ice-sheets. Major changes not directly related to the extent of ice include changes in the extent of monsoons.

The model's results for 18 ka are influenced by the existence of the continental ice-sheets, especially on the northern continents. However, there have also been changes within the Holocene as a result primarily of changing orbital parameters interacting with waning ice-sheets, especially in northeastern North America (see Fig. 3.6). Temperature departures were significantly lower in the southern continents for January and significantly higher in the northern continents for July, reflecting greater insolation in the northern hemisphere summer as summer solstice took place near to perihelion (see Fig. 3.3), while southern summers were cooler because their summer solstice occurred near to aphelion. Precipitation differences are evident in the monsoon areas of east Africa and southeast Asia (see also Kutzbach 1981; Kutzbach & Otto-Bliesner 1982).

Areas of near-zero departures in both temperature and precipitation differ between the 18 ka and 9 ka reconstructions. Nowhere on Earth has been immune from climatic changes of some magnitude. The global climate system must be seen as being in a state of continual flux.

The CCM and similar models are also being used to explore climates of the Earth with different continental configurations than occurred during the Quaternary. Using different models, Kutzbach & Gallimore (1989) and Crowley *et al.* (1989) investigated the climate of the Pangaean 'megacontinent' of about 250 Ma (Permian) using appropriate boundary conditions, and varying key conditions for which information is inadequate. In all experiments (Fig. 3.7), this large equatorial continent was predicted to have had an extreme continental climate, with hot summers, cold winters, vigorous monsoon circulation, and extreme seasonal contrasts, corresponding qualitatively with geological evidence for heat and aridity (Crowley *et al.* 1989; Kutzbach & Gallimore 1989).

Because of uncertainty about values for orbital parameters in pre-Quaternary periods, modelling experimenters have to make assumptions about what values to use. For example, for their investigations of Pangaea, Kutzbach & Gallimore (1989) assumed the modern value for obliquity but zero eccentricity, so that the model had no difference in insolation between the two halves of the year, and no precession effect. It is, of course, possible to experiment with possible values for the orbital

Climate models · 57

A   July temperature (°C)

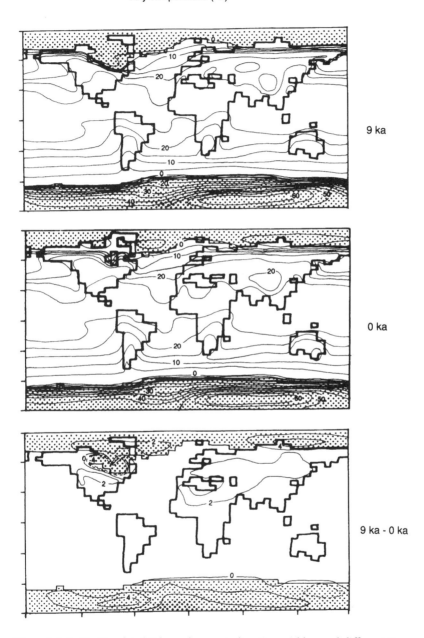

Figure 3.6 A–D. Simulated 9 ka and present climatic variables, and differences between them from GCM modelling experiments. Contours on the difference diagrams are dashed for negative values (present values greater than 9 ka values). Stippled area indicates extent of continental ice-sheets and floating ice. Continental landmasses are outlined in heavy, continuous lines. Redrawn from Kutzbach *et al.* (1993, Figs 4.1–4.4 and 4.9–4.12).

58 · Orbital-forcing of climatic oscillations

B

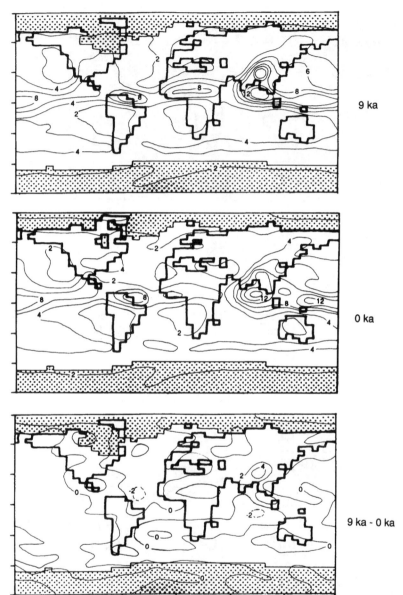

July precipitation (mm day⁻¹)

Figure 3.6 (continued) For legend see page 57.

Climate models · 59

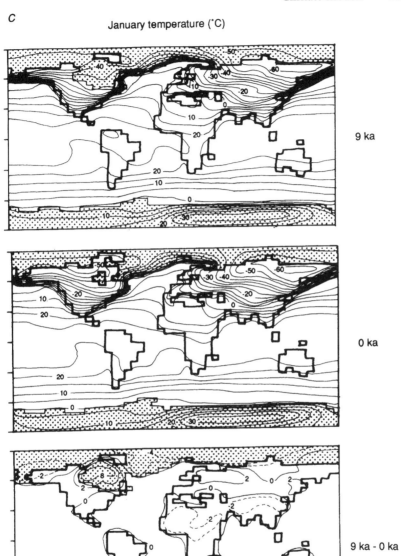

Figure 3.6 (continued) For legend see page 57.

60 · Orbital-forcing of climatic oscillations

D

January precipitation (mm day^{-1})

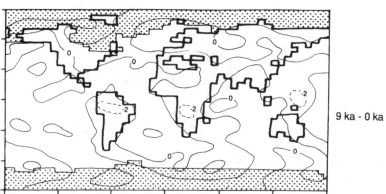

9 ka

0 ka

9 ka - 0 ka

Figure 3.6 (continued) For legend see page 57.

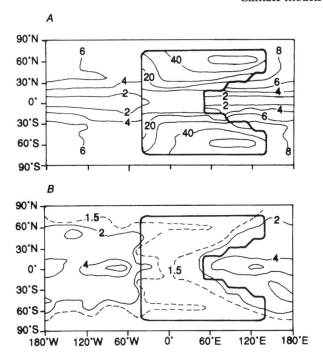

Figure 3.7 Predicted climates of Pangaea. Continental landmasses are outlined in heavy, continuous lines.
A. Surface temperature range (deg. Kelvin). B. Annual precipitation (mm day^{-1}). The temperature range over land is for summer minus winter, and over ocean it is for autumn minus spring.
Redrawn from Kutzbach & Gallimore (1989, Fig. 6).

parameters to discover the sensitivity of model results to them. For example, it has been suggested that the record of high-latitude Cretaceous and early Tertiary floras can be interpreted as indicative of warmth caused by reduced obliquity, and Barron (1984) described a CCM experiment for the Cretaceous designed to test this hypothesis. Reduced obliquity would increase high-latitude winter insolation, but reduce the amplitude of the seasonal oscillation and reduce mean annual insolation. Barron (1984) found that a reduction of obliquity from 24° (the modern value) to 15° is sufficient to produce substantial polar cooling to an extent incompatible with geological evidence for warmth. Another experiment with the CCM suggested that Cretaceous geography may have been responsible for global warming, especially in polar regions (Barron & Washington 1984).

Glancy et al. (1986) described a model experiment on the effect of the amplitude of orbital variations on the 20–100-kyr scale combined with Cretaceous continental configuration. In the absence of any information on the amplitudes and periods of Cretaceous orbital forcing, they assumed that Quaternary values would be a reasonable first approximation, and considered two extreme situations, named with reference to the northern hemisphere summer: MAX, which had eccentricity of 0.05, obliquity of 25°, and perihelion on 15 July; and MIN, which had eccentricity of 0.01, obliquity of 22°, and perihelion on 15 January. Experiments were run for northern hemisphere summer and winter, each with both MAX and MIN. MAX forcing produced surface summer temperatures up to 17 K warmer over northern continental interiors than MIN forcing (Fig. 3.8). In January, MAX forcing produced precipitation increases of over 6 mm day^{-1} (monsoon levels) on the coasts of southern North America and southeast Asia, while MIN forcing produced even greater increases on the coasts of modern northern South America and eastern Africa (Fig. 3.8). A surprise from the model was the prediction of extreme arid conditions over North America during the summer under MAX forcing, which may be due to some condition in the modelling simulation. However, Glancy et al. (1986) concluded that some areas of the globe show substantial climatic shifts between extremes of orbital circumstances. More modelling experiments with a variety of palaeogeographies and a range of orbital parameters are needed before it will be possible to understand the probable consequences of pre-Quaternary oscillations for climates at the Earth's surface and hence for the organisms that lived there.

The possible behaviour of Cretaceous climates in response to orbital variations is explored further by Park & Oglesby (1991), also using the NCAR CCM. They used the mid-Cretaceous palaeogeography and palaeotopography of Barron & Washington (1984), and orbital variations over the last 10 Myr (Berger 1978a) to approximate a reasonable range for the Cretaceous. Fig. 3.9 illustrates their results. In most cases, the model gave a greater response to precessional insolation changes than to obliquity changes, even at high latitudes. Changes in the model's evaporation–precipitation balance over the proto-South Atlantic suggested that a shift would occur from lagoonal to estuarine circulation in phase with the orbital fluctuations, and that this would be a viable mechanism for producing the lithological features found in cores of ocean sediment from the area.

The results of models are, of course, only as good as the models and

Climate models · 63

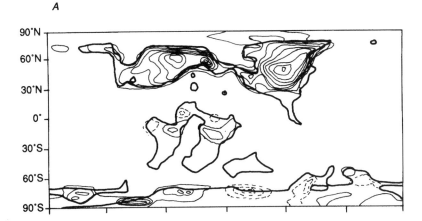

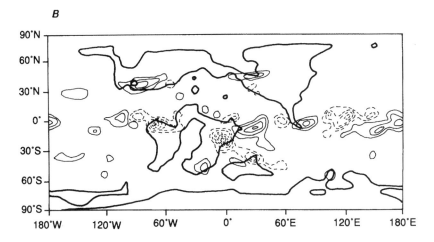

Figure 3.8 Predicted variation of climates with Cretaceous continental configuration. Continental landmasses are outlined in heavy, continuous lines. Values plotted are differences between those obtained using two extreme sets of orbital parameters (MAX and MIN: see page 62).
A. July temperatures (contour interval of 2 K).
B. January precipitation (contour interval 2 mm day^{-1}).
In both cases, continuous contours indicate higher values with MAX forcing, and dashed contours indicate higher values with MIN forcing.
Redrawn from Glancy *et al.* (1986, Figs 5 and 10).

64 · Orbital-forcing of climatic oscillations

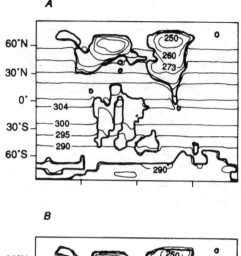

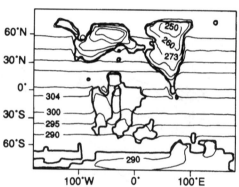

Figure 3.9 January surface temperatures (deg. Kelvin) for a 100-Ma (Cretaceous) global climate reconstruction for two orbital configurations.
A. Eccentricity factor = 1.068, and obliquity angle = 22.5°.
B. Eccentricity factor = 0.938, and obliquity angle = 24.0°.
The eccentricity factor is a term that includes eccentricity and precession as a measure of received solar insolation values. Modern eccentricity factor = 1.0. Redrawn from Park & Oglesby (1991, Fig. 3).

their inputs, and should not be taken too literally (Shackleton 1993a). Future models are likely to include a wider range of interactive elements, such as ice-sheets, oceans, and cloud cover. However, they do already illustrate the kind of global climatic changes that probably did take place because of the Earth's orbital variations and consequent insolation changes.

4 · Geological evidence for orbital-forcing

The recognition of rhythmic patterns in sediments with the same periodicity as orbital variations is a key part of the acceptance of orbital variations as causes of climatic changes at the Earth's surface. The initial impetus to investigate orbital variations was the problem of the Quaternary ice-ages: what caused them, and why the ice came and went (Imbrie & Imbrie 1979). To this extent, investigation of orbital changes was initiated by a particular, apparently rhythmic, series of geological events. The situation has now shifted to serious consideration of the other parts of the geological record because of the perpetual nature of orbital variations. This chapter reviews, in reverse stratigraphical order, evidence for the existence of rhythmic patterns throughout the geological record (see Table 1.1 for an outline time-scale), the origin of ice-ages, and concludes with a discussion of the nature of the record itself.

The evidence is mostly derived from two different sedimentary situations. There are laminated sediments, from lake and marine basins, that are grouped into units that recur rhythmically. Periodicity of such sediments is determined either by assuming (or demonstrating) that the laminae are annual and counting the laminae in each unit, by estimating the age of the whole sequence from radiometric dating and assuming that each unit is an equal representation of the whole time-span, or by applying a time-series technique (Schwarzacher 1993). The second situation is deep-sea sediment that is still accumulating, and from which a variety of chemical and isotopic values indicate periodic climatic conditions. These are the sediments that lead to the acceptance of orbital variations as the 'pacemaker' of the ice-ages (Hays, Imbrie & Shackleton 1976), and the main source of the geological evidence for orbital forcing in the Quaternary. However, a continuous 500-kyr record of $\delta^{18}O$ variations in a 36-cm core of vein calcite from Devils Hole, Nevada shows variations that, it has been claimed, are inconsistent with the Milankovitch hypothesis (Winograd et al. 1992; Ludwig et al. 1992). These claims have themselves been challenged, with respect to technical aspects of the dating (Edwards & Gallup 1993; Shackleton 1993b; see Ludwig et al. 1993 for reply), and because the dating of the vein calcite, if correct, suggests improbable accumulation rates for deep-sea sediment (Imbrie, Mix & Martinson 1993). Both records show, clearly, oscillations of climate at periodicities on the Milankovitch time-scales (Imbrie, Mix & Martinson 1993). Differences between them relate to the chronology of events, and the length of glacial and interglacial stages. Debate about the relative merits or otherwise of the two chronologies is likely to rage for some while. Crowley (1994) suggested that differences between the vein calcite chronology and the deep-sea chronology of Quaternary events may result from differing responses of the two systems to climate shifts associated with extensive glaciation. Climate systems are complex, and strong non-linear feedbacks are involved between climate, ocean, and ice-sheets, as well as orbital forcing (Ghil 1994).

Examples of sedimentary sequences that exhibit rhythmic patterns of accumulation on the Milankovitch time-scales are described below, beginning with the original Quaternary demonstration. Additional examples are reviewed by van Houten (1986) and Schwarzacher (1993).

Cenozoic

Investigation of the geological record of the Earth's orbital variations comes predominantly from the sediments in the Earth's oceans, which in some areas have been accumulating continuously through much of the Cenozoic to the present day. The record in these sediments has been pre-eminent in establishing the relationship between Quaternary climatic change and the Earth's orbital changes, and influential in the search for sedimentary oscillations in pre-Quaternary deposits. The single most significant piece of research in making a link that convinced most scientists concerned a detailed comparison between aspects of cores of ocean sediment and a model of the Earth's orbital variations by Hays, Imbrie & Shackleton (1976). They set out to test the orbital change hypothesis, this being the only idea about the frequencies of major Quaternary glaciations that had been formulated in such a way as to make testable predictions, and to address the two principal obstacles that had, to that date, inhibited acceptance of the orbital change hypothesis. These were uncertainty about which aspects of orbital change should be considered significant, and an uncertain chronology (Hays, Imbrie & Shackleton 1976). Two cores, RC11-120 and E49-18, both from the southern Indian Ocean and in an area known to have sediment accumulating fast enough for preservation of the necessary temporal resolution, were selected for study. Geological data were obtained on:

(i) Oxygen isotopic composition (expressed as $\delta^{18}O$) of the tests of the foraminifera *Globigerina bulloides*, using established techniques, which provided a record of the waxing and waning of continental ice-sheets through the consequential changes in the isotopic composition of sea-water (Shackleton 1967; Shackleton & Opdyke 1973).

(ii) T_s, an estimate of summer sea-surface temperatures derived from statistical analysis of radiolarian assemblages.

(iii) Relative abundance of *Cycladophora davisiana*, a radiolarian not included in T_s, which appeared to reflect the structure of summer surface waters (Hays, Lozano, Shackleton & Irving 1976).

These geological time-series are shown in Fig. 4.1.

Because the $\delta^{18}O$ record in deep-sea sediments reflects global ice volume, for the most part (Chappell & Shackleton 1986; Labeyrie *et al.* 1987), it is globally synchronous. The record provides a standard chronology, limited by ocean mixing (about 1000 yr) and bioturbation (Shackleton & Opdyke 1973), using a stratigraphy initiated by Emiliani (1955). Core V28-238 from the equatorial Pacific had extended $\delta^{18}O$ stratigraphy down to the Brunhes–Matuyama magnetic reversal at 730 ka, thus providing a time-scale for the stratigraphy. This sequence has become 'the Rosetta stone of the ice ages' (Shackleton & Opdyke 1973; Shackleton 1989). Hays, Imbrie & Shackleton (1976) obtained radiocarbon ages for certain horizons of RC11-120 and E49-18, and obtained estimates of the age of particular stage boundaries from other $\delta^{18}O$ sequences, particularly V28-238. Uniform accumulation rates were assumed between horizons for which there was an independent age estimate. Hays, Imbrie & Shackleton (1976) carried out spectral analyses on both the geological data (Fig. 4.2) and orbital and insolation variations of the past 468 kyr using the calculations of Vernekar (1972). The spectra of the geological data produced three peaks. The dominant peak had a period estimated to be 94, 106, and 122 kyr from T_s, $\delta^{18}O$, and *Cycladophora davisiana* respectively. A secondary peak had a period of 40–43 kyr and a third peak was unimodal at a period of 23–24 kyr for T_s and *Cycladophora davisiana*, and bimodal with subpeaks at periods of 24 and 19.5 kyr for $\delta^{18}O$ (Hays, Imbrie & Shackleton 1976). Using their chronology, which was completely independent of the astronomical theory, they found that spectra obtained from the geological data matched spectra from orbital variations so closely that the match should be accepted as a non-random

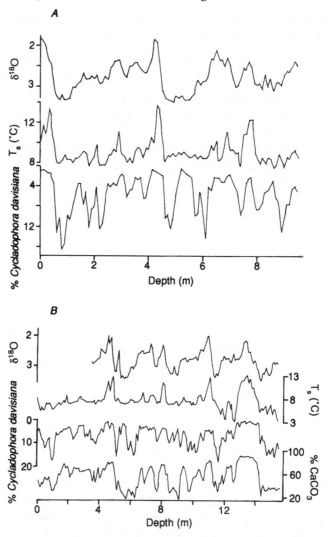

Figure 4.1 Geological time-series from cores of deep-sea sediment in the Southern Indian Ocean.
A. Data from core RC11-120.
B. Data from core E49-18.
Spectral analyses from these data are shown in Fig. 4.2.
Redrawn from Hays, Imbrie & Shackleton (1976, Figs 2 and 3).

Cenozoic · 69

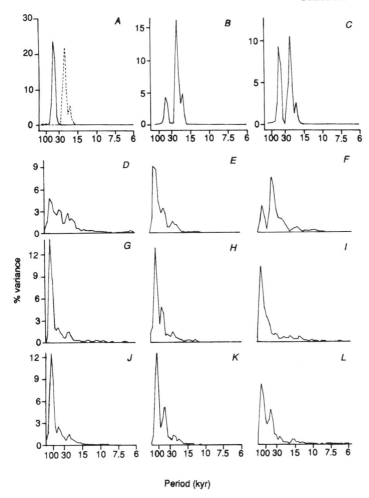

Figure 4.2 High-resolution spectra of orbital, insolation and geological variations. For each spectra, variance (as a percentage of total variance per unit frequency band) is plotted as a function of period.
A. Spectra for obliquity (continuous line) and precession (dashed line).
B. Spectrum for winter insolation at 55°S.
C. Spectrum for summer insolation at 60°N.
D–F. Geological spectra from core RC11-120 using the SIMPLEX age-depth model.
G–I. Geological spectra from core E49-18 using the SIMPLEX age-depth model.
J–L. Geological spectra of combined records using the ELBOW age-depth model.
Geological spectra are for reconstructed sea-surface temperature (D, G, J), $\delta^{18}O$ (E, H, K), and percentages of the radiolarian *Cycladophora davisiana* (F, I, L). See page 67, and Fig. 4.1 for additional details.
Redrawn from Hays, Imbrie & Shackleton (1976, Figs 4 and 5).

result. Modal frequencies in the geological record matched the obliquity and precession frequencies to within 5%, and included peaks for both that were significant at the $p = 0.05$ level, and predicted ratios of obliquity to precession frequencies matched the ratios in the geological data to within 5% (Hays, Imbrie & Shackleton 1976). One surprise that turned up was the dominance in the geological data of the 100-kyr period, close to the period of eccentricity variation. Although this had been noticed before, it was not clear how the eccentricity variations could be translated into insolation, and hence climate, variations. The amount of insolation received by the Earth over an entire year varies only slightly with eccentricity variation (about 0.1%), which seemed too small to be the dominant influence of global climatic change. The distribution of insolation by season is strongly influenced by eccentricity changes, and this may be the explanation (Imbrie & Imbrie 1979). Imbrie et al. (1993) suggested that the northern hemisphere ice-sheets, after reaching a certain size, drive atmospheric and oceanic responses in a way that mimics external forcing, and this may be responsible for the dominance of the 100-kyr peak in the middle and late-Quaternary.

Hays, Imbrie & Shackleton (1976) continued their analysis by smoothing the data with filters and then determining, for each frequency component, the position of inflection points in the filtered record. The three sets of geological data, filtered at 40 kyr, are approximately in phase throughout their entire length. The $\delta^{18}O$ record lags T_s by 2 kyr, and *Cycladophora davisiana* by about 1 kyr. Data filtered at 23 kyr show $\delta^{18}O$ in phase with *Cycladophora davisiana* throughout the record, but lagging by about 4000 yr, while $\delta^{18}O$ systematically lags T_s after 350 kyr by about 4000 yr, but before then $\delta^{18}O$ and T_s were out of phase. Thus, climate changes of the two hemispheres are roughly in phase, with changes in southern hemisphere oceans leading changes in northern hemisphere ice-sheets by only a few thousand years. Comparisons between geological data and orbital variations over the last 150 kyr (when the chronology is most certain), show changes in $\delta^{18}O$, T_s, and *Cycladophora davisiana* lagging precession by about 3000 yr. Times of low temperature, high $\delta^{18}O$ (hence high continental ice volume) and abundant *Cycladophora davisiana* are associated with times of larger than average summer Earth–Sun distance.

These analyses present strong evidence for orbital control of major Quaternary climatic changes. Irregularities that occurred in the relationship between geological and orbital variations were concentrated in the early part of the record, and Hays, Imbrie & Shackleton (1976) were

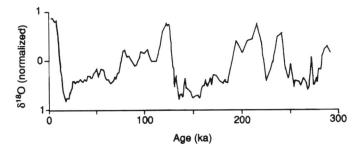

Figure 4.3 Orbitally-based chronostratigraphy for the late-Quaternary. Peaks indicate periods of minimum glaciation (for example the present interglacial, since 10 ka, and the last interglacial, at about 125 ka). Troughs, correspondingly, indicate glacial maxima. Redrawn from Martinson *et al.* (1987, Fig. 18).

able to remove the irregularities by making a small adjustment to the time-scale (but within the limits of the radiometric time-scale), and thus extend the phase relationships back for the whole extent of the record. Their paper marked the beginning of general acceptance of the hypothesis of orbital control of the pace of Quaternary ice-ages. Subsequent work built on that acceptance by using orbital variations to 'tune' the geological time-scale. This is done by assuming that the two records should be in phase, but that only the calculated record of insolation change has an accurate chronology. Thus, the insolation chronology can be used to compensate for variations in sediment accumulation rate. The $\delta^{18}O$ isotope record of deep-sea cores forms the most complete record of Quaternary climatic changes, now being extended back into the Pliocene (N.J. Shackleton, cited in Harland *et al.* 1990), with calculated orbital variations increasingly providing the key to chronology (for example Martinson *et al.* 1987: see Fig. 4.3).

During the late-Quaternary, the 100-kyr period for eccentricity variation dominates the geological record, but during the early Quaternary, the 41-kyr obliquity period dominates the record (Ruddiman *et al.* 1986; Ruddiman & Raymo 1988; Shackleton *et al.* 1988). Before 2.4 Ma, evidence from variations of terrigenous material in deep-sea sediments suggests that the 23- and 19-kyr periodicities may then have dominated in the tropics (Bloemendal & deMenocal 1989). Thus, interaction of the variation in the various orbital parameters, and variation in their effects with latitude, combine to produce a geological record whose dominant periodicity changes with time.

A continuous 850-m core of Cenozoic sediment with periodic alterations of clay-rich and clay-poor beds has been obtained from Deep Sea Drilling Project Site 336 off northwest Africa (Dean et al. 1981). The sedimentary variation has developed as a consequence of carbonate dissolution brought about by shallowing of the carbonate compensation depth and climatically-induced fluctuations in the thickness of Antarctic bottom water. Using a time-scale based on sediment accumulation rates, it was estimated that carbonate-dissolution oscillations have periods of about 44 kyr in the early Miocene and Oligocene, 19 kyr in the middle and late Eocene, and 7 kyr in the early Eocene. Dean et al. (1981) concluded that these sediments are recording responses to orbital variations just as in the Quaternary climatic record, but with dominant periodicities varied because of other influences. They suggested that climatic interpretations of the Quaternary record can therefore be extended back at least into the Oligocene, and possibly the Eocene. Moore et al. (1982) and Tiwari (1987) have found much longer periods in Miocene deep-sea sediments. Spectral analyses of $\delta^{18}O$ and $\delta^{13}C$ records in a core from Deep Sea Drilling Project Site 518 in the south Pacific Ocean showed periodicities of 2.0 and 1.25 Myr and 800, 400, 115 and 95 kyr, matching eccentricity variations but with other, longer period, oscillations (Tiwari 1987). High-precision analysis of another deep-sea Miocene sequence revealed a periodicity of 40 kyr in the $\delta^{18}O$ record from 16.2–16.8 Ma (Pisias et al. 1985).

The climatic variance in $\delta^{18}O$ records from deep-sea cores spanning 130 Myr at temporal resolutions of 1 kyr to 10 Myr, plotted as a composite variance spectrum (Fig. 4.4) shows broad concentrations of variance on the Milankovitch time-scales (periods of 20–100 kyr), and a concentration near a period of 30 Myr (Shackleton & Imbrie 1990). There is also a clear variance minimum between about 200 kyr and 3 Myr. The variance concentration near a period of 30 Myr may be related to the periodicities claimed for mass extinctions seen in the fossil record of marine invertebrates (Raup & Sepkoski 1984, 1986, 1988; Sepkoski 1989). This is clear evidence that variation in climate can be expected on the Milankovitch time-scales at least throughout the Cenozoic, and that variation near a period of 30 Myr is probably the only other source of climatic change that need be considered at periodicities below those related to continental formation and break-up at periods near 400 Myr.

A series of cores from deep-sea sediments in the South Atlantic, spanning the Cretaceous–Tertiary boundary, display colour variations due to fluctuating abundances of calcium carbonate. Analysis of these by

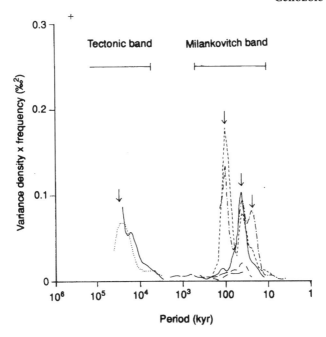

Figure 4.4 Composite variance spectrum for seven $\delta^{18}O$ datasets. A cross (top left of plot) gives an estimate of the 400 Myr trend. Redrawn from Shackleton & Imbrie (1990, Fig. 4).

quantifying the optical transmission of photographs of the cores has enabled spectral analysis of the fluctuations for sediments spanning 20 Myr (Herbert & D'Hondt 1990). Time-scales for the cores were provided by palaeomagnetic reversals, and revealed that the carbonate fluctuations have periodicities averaging 23.5 ± 4.4 kyr (Fig. 4.5).

Terrestrial records with evidence of periodicity of climatic changes will be discussed in Chapter 5, with discussion of the effects of Quaternary climatic changes on the distribution and abundances of organisms. There are few Tertiary sequences that have been sampled finely enough to yield any information about rates of climatic change or biological responses. One exception is the Teewinot Formation, a 2000-m thick outcrop of late Miocene lacustrine sediments in Wyoming, USA, studied palynologically by Barnosky (1984). Laminated sediments, and estimated accumulation rates based on observed values in similar modern systems, suggest that the upper 68 m accumulated in 30–300 kyr at about 9.2 Ma. From the pollen record for this section (Fig. 4.6), Barnosky (1984)

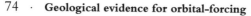
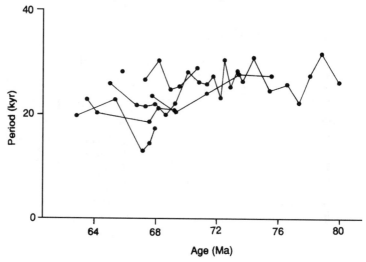

Figure 4.5 Periodicity of carbonate oscillations as revealed by optical transmission of deep-sea sediment cores spanning the Cretaceous–Tertiary transition in the South Atlantic. Each point is the mean value for a 9.5-m core length. Points from the same core site are connected by lines in stratigraphic sequence. Drawn from data in Herbert & D'Hondt (1990, Table 1).

infered a vegetation with a high proportion of conifers throughout, but with fluctuating abundances of *Sarcobatus* (a halophytic shrub), Cupressaceae, *Ephedra*, deciduous hardwoods ('Tertiary relicts'), and herbs. The vegetation composition around this site has shifted on a time-scale of 1–200 kyr (Barnosky 1984).

Mesozoic

Cretaceous rhythmic bedding sequences have attracted particular attention, beginning with Gilbert (1895), and producing reviews by Barron *et al.* (1985) and the Research on Cretaceous Cycles Group (1986). Herbert & Fischer (1986) described especially high-resolution data on redox conditions and carbonate rhythmicity from deep-sea Albian (mid-Cretaceous) limestones of central Italy. Calcium carbonate and redox state (using light transmission on transects of diapositives) were measured on an 8-m section of core containing about 74 couplets grouped in about 17 bundles. The couplets consist of calcareous, oxidized beds alternating with shaly, reduced beds, due to variations in biogenic carbonate

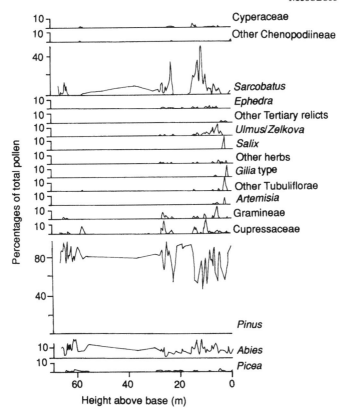

Figure 4.6 Abundance of pollen types in the Miocene sequence of lake sediments in Wyoming, USA. Redrawn from Barnosky (1984, Fig. 2).

productivity, and the bundles appear as a consequence of lower frequency variation in carbonate content and colour (Fig. 4.7). The thickness of Albian sediments in the core is 65 m, accumulated at an average rate of 5.0 m Myr^{-1}, yielding a couplet period of 21–22 kyr, and a bundle period of 95–96 kyr. The more recent geological time-scale of Harland *et al.* (1990) gives the duration of the Albian as 15 Myr, making the periods about 25 and 109 kyr, respectively. There is also a long oscillation, with a period of about 400 kyr. Herbert & Fischer (1986) continued the analysis by comparing their Albian data with 1500 kyr of recent eccentricity variation, and undertaking spectral analysis. Virtually all of the sedimentary variance lay within Milankovitch time-scales, and there is a general resemblance to the Quaternary δ^{18}O spectrum in that most

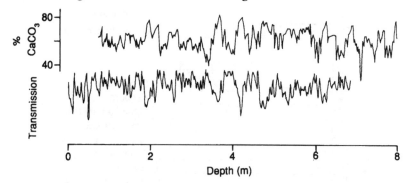

Figure 4.7 Redox and carbonate content for part of an Upper Albian (mid-Cretaceous) sequence of central Italy. Values for transmission (brightness), used as a measure of redox state, increase downwards, mirroring carbonate data. Sediment accumulation rate is about 5 m Myr^{-1}. Redrawn from Herbert & Fischer (1986, Fig. 1).

of the amplitude is concentrated in the peak near 100 kyr (Fig. 4.8). Higher frequency variation, such as the couplet frequency, is masked by the variability in sedimentation rates at this fine temporal scale (see also Park & Herbert 1987). These periodicities suggest the existence of a dynamic ocean–climate system, in which orbital variations forced wide swings in the climate regime. Since the Cretaceous seems to have been a time during which there was, globally, negligible glacier development (see page 85), these data indicate that ice-sheet amplification may not be a prerequisite for large-scale climate modulation by periodic variations in insolation (Herbert & Fischer 1986).

Barron *et al.* (1985) described the bedding rhythms of the mid-Cretaceous Greenhorn Formation on the northern edge of the Tethyan Ocean in North America. Biota (trace fossils, shelly macrofauna, encrusting organisms, and calcareous microfossils), sedimentology, and geochemistry suggested variations in precipitation, runoff and stability of marine stratification. They argued that results of their computer modelling indicated that the Tethyan Ocean, as a large zonal ocean in the subtropics, exerted a dominant influence on general circulation. Its northern margin would have been characterized by intense winter precipitation, and may have been highly sensitive to orbital variations. The model experiments of Glancy *et al.* (1986) support this interpretation (see page 62).

Rhythmicity of sedimentation is readily apparent in the Upper Cretaceous chalk sequence of southern England (Hart 1987). It is usually

Mesozoic · 77

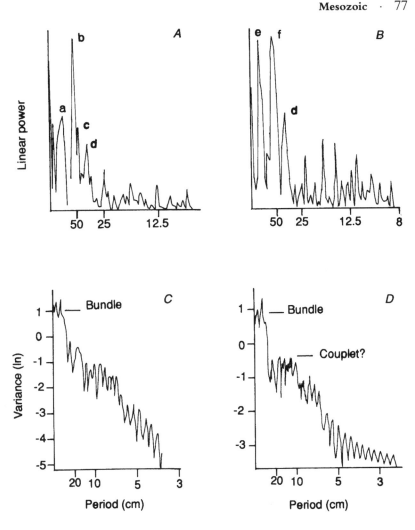

Figure 4.8 Transmission and carbonate power spectra from part of an Upper Albian (mid-Cretaceous) sequence of central Italy, based on data in Fig. 4.7.
A. Carbonate power spectrum showing prominent peaks at 90 cm or 225 kyr (a), 53 cm or 108 kyr (b), 48 cm or 98 kyr (c), and 36 cm or 73 kyr (d).
B. Transmission power spectrum showing prominent peaks at 171 cm or 347 kyr (e), 57 cm or 116 kyr (f), and 36 cm or 73 kyr (d), as in *A*.
C. Lower-resolution power spectrum of carbonate plotted as ln variance.
D. Lower-resolution power spectrum of transmission plotted as ln variance.
Redrawn from Herbert & Fischer (1986, Fig. 3).

seen as marl–limestone alternations, fining-upward on a 0.2–3.0 m scale, with individual units traceable over tens of kilometres (Robinson 1986; Hart 1987). Using the number of microrhythms per stage (Robinson 1986) and the geological time-scale of Harland *et al.* (1990), the periodicity of rhythms in the Upper Cretaceous is 47 kyr (Cenomanian Stage), 18 kyr (Turonian Stage), 30 kyr (Coniacian Stage), and 90 kyr (Santonian Stage).

Basal Jurassic sediments in southern Britain accumulated under a shelf regime, and consist of carbonate mud from coccolithophores and riverine clays (Weedon 1986). The pattern of accumulation was rhythmic because of changes in bottom-water oxygenation and changes in the clay : carbonate mud ratio. Spectral analyses revealed two periodicities, with a ratio of 1.6 between them. Weedon (1986) argued that these may be equivalent to Jurassic obliquity and precession oscillations in the range 20–100 kyr.

The Triassic geology of the southern Alps in Italy is characterized by shallow water platform carbonates, consisting of hundreds of metre-scale oscillations, each composed of subtidal sediments capped by sediments considered to be subaerially-weathered sediment (Hardie *et al.* 1986). These sediments were interpreted as the result of sea-level oscillations (Fig. 4.9), and the oscillations were calculated to have had an average duration of a few tens of thousands of years, based on overall division ages for the stages concerned. Hardie *et al.* (1986) noted the absence of peritidal sediments between the subtidal sediments and the diagenetic caps of each oscillation, ruling out the possibility that the subaerial exposure was due to shoreline progradation. The repeated subaerial exposure must, therefore be due to sea-level oscillations of either eustatic or tectonic origin, or some combination. They argued that continuous tectonic 'yo-yoing' is rather unlikely, although not impossible, so the oscillations probably have a glacio-eustatic origin, linked with climatic oscillations at Milankovitch time-scales (Hardie *et al.* 1986; Goldhammer *et al.* 1987).

During the Triassic, a sequence of annually-laminated sediments formed in a series of rift valley lakes in what is now eastern North America (Olsen *et al.* 1982; Olsen 1984, 1986, following van Houten 1962). The lakes covered more than 7000 km^2 and were over 100 m deep, but were reduced, or dried out completely, during low stands, and thus fluctuated with an amplitude similar to that of large Quaternary lakes in western North America or east Africa (Olsen 1984). The longest continuous lacustrine record (Fig. 4.10) consists of a spectrum of sedimentary os-

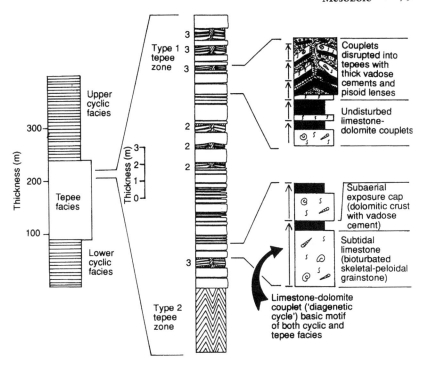

Figure 4.9 Vertical stratigraphic arrangement of facies in the Ladinian Latemar Limestone, from the Triassic of northern Italy. On the right is a typical segment of the detailed measured section of the tepee facies showing a succession of thin limestone–dolomite couplets interrupted at intervals by seven Type 1 tepee zones (each involving two or three couplets) and a Type 2 tepee zone. The cyclic facies successions differ from the tepee facies only in the lack of tepee zones. Reproduced from Hardie et al. (1986, Fig. 2).

cillations, averaging about 5 m in thickness, due to the rise and fall of the lake level. Olsen (1984) found that individual oscillations can be uniquely defined by their content of fossil fishes and reptiles, and are traceable over many tens of kilometres. As well as the 5-m thick beds, there are also oscillations of about 25 m and 100 m in the nature of sedimentation within each short oscillation. Portions of the 5-m thick beds are microlaminated and, assuming that these laminae are annual, each short oscillation lasted an average of about 21 kyr. The other oscillations are about 101 kyr and 418 kyr. Estimation of the periodicity from the published radiometric time-scale of the Triassic suggested periods of 24, 119 and 476 kyr (Olsen 1984), and Fourier analysis suggested periods

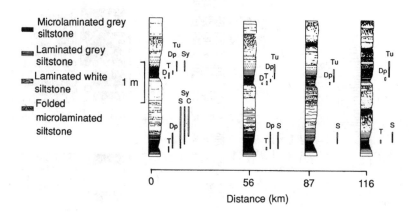

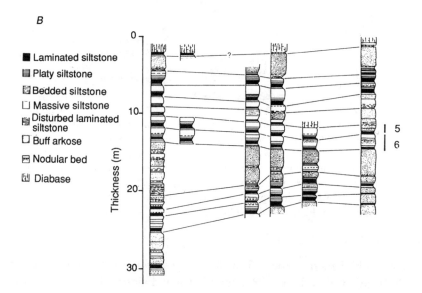

Figure 4.10 Lateral extent of detrital cycles from the Lockatong formation of the Triassic, New Jersey, USA.

A. Detailed correlation by vertical distribution of fossil vertebrates of cycles 5 and 6, marked on *B*.

B. Lateral correlation of all exposed cycles.

Abbreviations for vertebrates are D, 'Deep-tailed swimmer'; T, *Tanytrachelus*; Dp, *Diplurus*; Tu, *Turseodus*; Sy, *Synorichthys*; S, *Semionotus*; C, *Cionichthys*. Reproduced from Olsen (1984, Fig. 5).

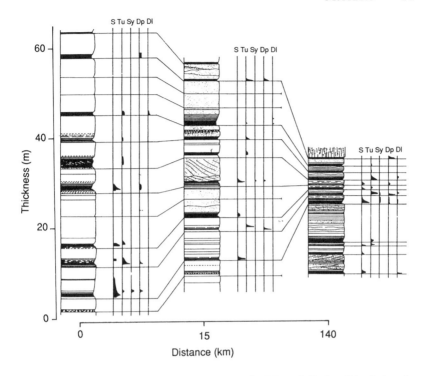

Figure 4.11 Correlation of three sections across the Newark Basin of detrital cycles of the Weehawken member of the lower Lockatong formation of the Triassic, northeastern USA, showing the distribution of the most abundant fish. Abbreviations for fishes are S, *Semionotus*; Tu, *Turseodus*; Sy, *Synorichthys*; Dp, *Diplurus* sp.; Dl, *Diplurus longicaudatus*. Reproduced from Olsen (1984, Fig. 6).

of 25, 44, 100, 133 and 400 kyr (Olsen 1986). The palaeolatitude of the lakes was about 15°N, where obliquity effects would be expected to be minimal, and the amplitude of oscillations at the modern obliquity frequency is weaker than the other oscillations (Olsen 1986). It seems that climatic changes, particularly precipitation and evaporation, due to orbital variations produced substantial changes in the environments of these lakes, passing from deep-water to playas in a few tens of thousands of years. This must have had dramatic ecological effects, which is probably why the vertebrate fauna varies from one short oscillation to another (Fig. 4.11: see also page 157).

Paleozoic

A sea-level curve for the Upper Carboniferous (Pennsylvanian) in North America has been produced by Heckel (1986), based on a sequence of at least 55 oscillations of transgression and regression (Fig. 4.12). These sea-level changes affected much of what is now mid-continental North America. Estimates of the duration of the oscillations range from 44–120 kyr for minor oscillations (less extensive geographically) up to 235–400 kyr for major oscillations (more extensive geographically), providing empirical evidence for control of Carboniferous stratigraphic patterns by orbital forcing. Heckel (1986) suggested that understanding the Quaternary may well be the key to understanding the Carboniferous because of the similarity of periods of orbital forcing at a time of continental glaciation.

An 1100-m succession of Upper Devonian age in east Greenland, consists of meandering river and associated overbank deposits, with sandstone bodies interpreted as point bars (Olsen 1990). Periodic variations in discharge, channel slope and channel, about 20 m thick and grouped into 100 m 'megacycles', were found. Using the published radiometric time-scale, Olsen (1990) estimated an average accumulation rate for the sequence of about 0.5 mm yr^{-1}, giving estimates of 42.3 and 233 kyr for the periodicity of the oscillations. Olsen (1990) considered that the changes were an expected result of precipitation changes, forced by orbital variations, whereas the alternative (tectonism) would have had different consequences for the development of the river channels.

In western Texas and southeastern New Mexico, a deep sediment-starved basin filled during the Permian with a 600-m sequence of evaporites (Anderson 1982, 1984). The sediment consists mostly of thin laminae of calcium sulphate alternating with thinner laminae of dark brown organic-rich calcite in couplets 2–4 mm thick that can be traced over 14 km (Anderson 1984). The relative thicknesses of the two halves of the couplet vary within the sequence in periodic fashion, with dominant periods (assuming that the couplets represent annual accumulation) of 2.7, 20, and 100 kyr (Fig. 4.13). The oscillations cannot be attributed to progressive changes within the basin, so Anderson (1982, 1984) ascribed them to climatic changes forced by orbital changes of eccentricity and precession, with obliquity missing because of the low palaeolatitude.

The Mallowa Salt in Western Australia is an evaporite sequence of late Ordovician to early Silurian age in coastal and ephemeral salt pan and saline mudflat environments. Geochemical analysis of a 477-m core

Paleozoic · 83

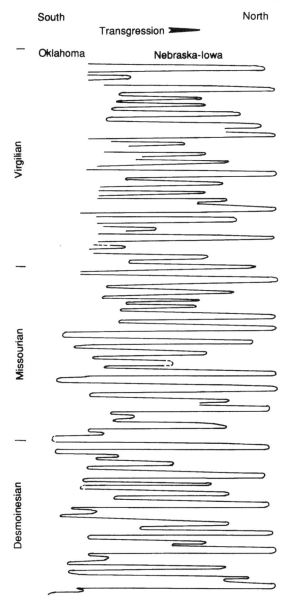

Figure 4.12 Sea-level curve for part of the Middle–Upper Pennsylvanian (Carboniferous) sequence along the North American mid-continent outcrop belt, based on shoreline positions estimated from the furthest basinward extent of exposure surfaces and fluviodeltaic complexes, and the furthest shelfward extent of marine horizons or deepest water facies at the northern outcrop limit. Current age estimates for the time-scale are 303 Ma for the Desmoinesian–Missourian boundary, and 290 Ma for the top of the Virgilian (Harland *et al.* 1990). Redrawn from Heckel (1986, Fig. 2).

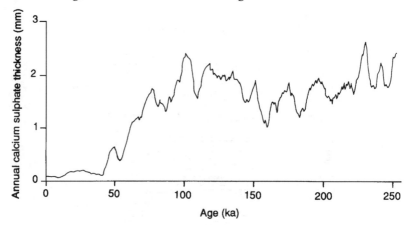

Figure 4.13 Smoothed plot of absolute thickness of calcium sulphate in the Castile Formation, from the Permian of southern USA. Smoothing interval is 8 kyr. Age is measured from the base of the sequence. Redrawn from Anderson (1984, Fig. 3).

displayed conspicuous periodicity on several orders, which Williams (1991) analysed by spectral analysis (Fig. 4.14). Period ratios of the major peaks are, to within one standard deviation of the core period measurement, similar to period ratios for the spectral peaks of orbital eccentricity and precession variations, after allowing for likely changes in the Earth–Moon system since 440 Ma (Berger *et al.* 1989a, b). Relative amplitudes of the major peaks are also similar to spectral peaks of orbital variations (compare Fig. 4.14 with Figs 4.2 and 4.4). The lack of any obliquity peak in the Mallowa Salt record is consistent with the known low palaeolatitude of deposition. The study indicates that the content of some elements in halite evaporite sequences may provide a sensitive record of climatic oscillations. They also indicate that there are variations in depositional environments as habitats for organisms on the same time-scale.

Proterozoic

The detail available in the geological record diminshes with distance back from the present, and rocks of Proterozoic age and older offer generally coarse stratigraphic resolution. Even here, however, there are glimpses of fine detail in unmetamorphosed sediment sequences. Periodic shelf sedimentation of early Proterozoic age (1.9 Ga) has been described in northern Canada by Grotzinger (1986a, b). A variety of facies occur on slope, outer shelf, shoal complex, inner shelf, and lagoon as sea-level rose and fell rhythmically with an amplitude of about 10 m (Fig. 4.15). From the duration of deposition of the formation, and the number of oscillations present, he calculated a period of 18–30 kyr, but possibly up to 90 kyr, depending on the accuracy of the geochronology. Although the estimate has a large range, it provides evidence of oscillations in sea-level at Milankovitch time-scales, which implies that the sea-level fluctuation was being forced climatically, probably through the formation of mountain glaciers or local continental glaciation (Grotzinger 1986a, b). Grotzinger (1986a) reviewed the occurrence of other Proterozoic periodic platform sequences that have similar periodicities (measured as sediment thicknesses).

Continental glaciation in Earth history

Cenozoic data provide the strongest evidence yet that the Earth's orbital variations do translate into substantial climatic changes at the Earth's surface. The data presented above support the existence of oscillations within the Milankovitch time-scales continuously back from the Quaternary, with its known fluctuations of major continental ice-sheets, into and through the Tertiary. This still leaves open the question of the cause of glacial periods in Earth history, such as the Quaternary. Orbital variations have been established as the factors controlling the pace of climatic fluctuations, but they have been present throughout Earth history, so their existence does not explain why these fluctuations produce continental ice-sheets at some times during Earth history but not others (Table 4.1, p. 88). The causes of glaciation have been discussed in most detail with respect to the Quaternary, and the principal possibilities are discussed below.

Continental configuration

Cenozoic movement of continental plates has nearly enclosed the Arctic Ocean, and Antarctica has moved over the South Pole following the breakup of Gondwana, effectively blocking the movement of warmer ocean currents and air masses from equatorial regions, and thereby permitting the development of permanent ice-sheets. These, it is argued, are then self-perpetuating because of their high albedo and they have a general cooling effect on climates of adjacent regions. Continental ice-sheets then develop during orbitally-controlled times of minimum insolation (Imbrie & Imbrie 1979).

86 · Geological evidence for orbital-forcing

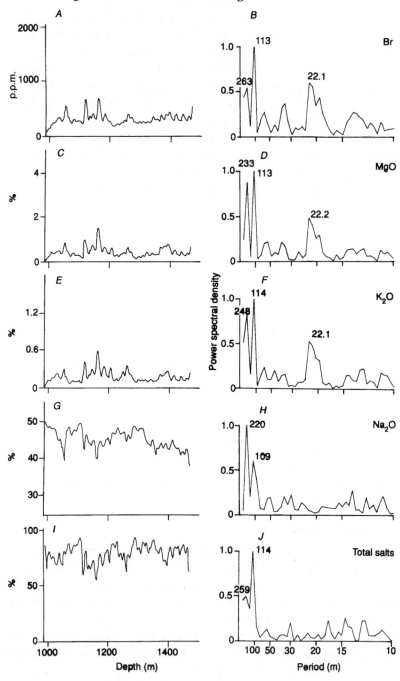

Figure 4.14 For legend see facing page.

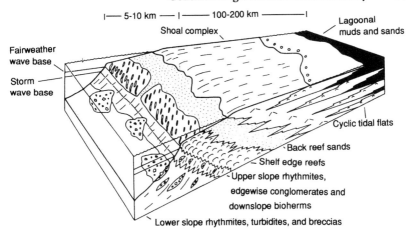

Figure 4.15 Shelf palaeogeography of the Rocknest formation from the Proterozoic of northern Canada. Cyclic deposits are formed by limited westward (to left on Figure) and extensive eastward progradation of the shoal complex in response to low-amplitude, high-frequency sea-level oscillations. Reproduced from Grotzinger (1986a, Fig. 3).

Tectonic uplift

Uplift of the plateaux of western North America and Tibet during the Cenozoic may have changed basic wind directions at all atmospheric levels with far-reaching climatic consequences for the climates of land and oceans (Ruddiman & Raymo 1988; Ruddiman et al. 1989; Ruddiman & Kutzbach 1990). Geological evidence suggests uplift of the Tibetan plateau by at least 2 km since 10 Ma, and of western North America by about 1 km during the last 15 Myr, with additional significant uplift in the Andes and New Zealand (Ruddiman et al. 1989). Kutzbach et al. (1989) carried out modelling experiments with the NCAR CCM to test this hypothesis by comparing model output using modern continental configurations with no mountains, half mountains and full (modern) mountains. They found that atmospheric heating rates, tropospheric

---

Figure 4.14 Salt content and Fast Fourier spectra from the geochemical stratigraphic series for the Mallowa Salt (late Ordovician – early Silurian of Western Australia). *A, C, E, G,* and *I* are plots of salt content, smoothed by a spline function. *B, D, F, H,* and *J* are Fast Fourier transform smoothed spectra. Peaks significant at the 95% level of confidence are labelled. Redrawn from Williams (1991, Figs 3 and 4).

Table 4.1. *Summary of known periods of pre-Quaternary glaciation during Earth history, based largely on till-like deposits*

Time interval	Glacial record
Tertiary (65–1.64 Ma)	Sea-level glaciers in Antarctica during Eocene–Oligocene time
Cretaceous (132–65 Ma)	No unequivocal evidence of glaciation
Jurassic (208–132 Ma)	No unequivocal evidence of glaciation
Triassic (245–208 Ma)	No unequivocal evidence of glaciation
Permian (290–245 Ma)	Extensive glaciation in Africa, Antarctica, southern Asia, Australia, and South America
Carboniferous (363–290 Ma)	Extensive glaciation in Africa, Antarctica, southern Asia, Australia, and South America
Devonian (409–363 Ma)	Glaciation in west Africa and, possibly, South America
Silurian (439–409 Ma)	Intensive glacial activity in South America
Ordovician (510–439 Ma)	Early Ordovician: local glaciation in northwest Europe Late Ordovician: extensive continental glaciation in north and west Africa, smaller ice-sheets in southern Africa and eastern North America
Cambrian (570–510)	Glaciation in west Africa and, possibly, China
Proterozoic (2500–570 Ma)	Early Proterozoic: extensive continental glaciation in North America, southern Africa, India, and Western Australia 1000 Ma – early Cambrian: repeated glaciation in all continents except Antarctica
Archaean (4000–2500 Ma)	Glaciation in southern Africa at about 2600 Ma

Note: There is also evidence for sea-ice activity in most periods, including the Mesozoic.
Source: From Hambrey & Harland (1981), using the time-scale given in Table 1.1.

motion and planetary wave amplitudes varied linearly with progressive uplift. These effects produced significant climatic changes supported by changes recorded in the geological record: winter cooling of the northern parts of northern hemisphere continents, summer drying of the west North American coast, interior Eurasia and the Mediterranean, winter drying of central North America and central Asia, and changes over the North Atlantic ocean likely to favour formation of northern source deep water (Ruddiman & Kutzbach 1989). These changes resulted from increased orographic diversion of westerly winds, from surface flows induced by summer heating and winter cooling of the uplifted plateaux, and intensification of atmospheric circulation due to exchanges between summer-heated plateaux and mid-latitude oceans. They amount to much of the late Cenozoic climatic deterioration that resulted in Quaternary ice-ages, and suggest that uplift was an important forcing factor for northern hemisphere climatic change at timescales longer than orbital variations (Ruddiman & Kutzbach 1989). It is likely that such tectonic changes are responsible for much of the variation in extent of glaciations over the course of Earth history (Eyles 1993).

However, GCM experiments suggest that tectonic uplift cannot by itself account for the general trend of global cooling observed to have taken place over the last 100 Myr. It may be that this cooling is related to a general lowering of the carbon dioxide levels in the atmosphere, but this may itself be related to uplift (Ruddiman 1990). Uplift produces steep slopes and alters atmospheric circulation in a way that strengthens monsoonal circulation. The higher rainfall on steeper slopes increases erosion, and the chemical weathering of fresh silicates can involve removal of carbon dioxide from the atmosphere.

Solar System dynamics

It has also been suggested that glacial periods in Earth history (see Table 1.1) are themselves periodic, and are due to the orbit of the solar system around a galactic centre (McCrea 1975; Williams 1975). The dynamic history of the Earth may be described as a consequence of a hierarchical system of time-delayed coupled oscillators (Shaw 1994). The lithostratigraphic and biostratigraphic record that forms the basis of the geological time-scale is, in effect, a non-linear clock that records the history of connected processes of Earth and Solar System dynamics. The pattern of appearances and disappearances of ice-ages, and individual glacial–interglacial periods, can thus be seen as consequences of the Earth's status within the Solar System, and to have an ultimate cause in Solar System dynamics as well as proximal causes in terms of Earth processes, whether tectonic, atmospheric, or both.

Discussion

Goodwin & Anderson (1985) and Anderson & Goodwin (1990) argued that virtually all of the stratigraphic record, throughout geological time, 'consists of thin, basin-wide shallowing-upward cycles bounded by surfaces produced by geologically instantaneous relative base-level rises (punctuational events)' (Goodwin & Anderson 1985, p. 515). They coined the acronym 'PAC' (Punctuated Aggradational Cycle) as a label for this new stratigraphic unit. PACs, they argued, are characteristically a few tens of thousands of years in duration. Goodwin & Anderson (1985) suggested that glacio-eustatic sea-level changes forced by orbital variations are the most likely explanation for the widespread nature of PACs, pointing out that there is no other likely mechanism in this frequency band that is allogenic, small-scale, episodic, and persistent throughout geological time.

Orbital variations have sufficiently large amplitude to force climatic changes continuously throughout geological time, causing climatic and base-level changes that produce the episodic stratigraphic units widespread in the stratigraphic record. It has been suggested that eventually it may be possible to obtain an absolute time-scale from the measurement of rhythmicities in the stratigraphic record (House 1985; Herbert & D'Hondt 1990). Regardless of whether the proposed allocycle (Anderson & Goodwin 1990) receives acceptance as a stratigraphic unit, the hypothesis about the way in which the sedimentary record accumulated has considerable support in the rock record (see above) and has profound implications. For example, changing environments of deposition every few tens of thousands of years implies shifting distributions of organisms that live in those environments at the same time-scale. Since these changes are global in extent, they would also involve terrestrial environments and the organisms that occupy them. This interpretation of the rock record thus compels consideration of the dynamicity of populations and communities of organisms on time-scales that are much finer than are usually considered in pre-Quaternary sequences. This topic will be developed in Chapter 8 after consideration of the effects of climatic oscillations at Milankovitch time-scales on organisms in the much better-studied Quaternary record.

The evidence presented in this chapter demonstrates the existence of sedimentary oscillations throughout the geological record. Where it has been possible to obtain rates of change, the periodicities of the oscillations fall within Milankovitch time-scales, and the most likely explanation for them is that they were forced by climatic oscillations resulting from the Earth's orbital variations. From the point of view of consequences for life on Earth, the fact that a pattern of repeated and constant change of environments exists is the important aspect, rather than the cause. But the geological record is far from perfect, and it might be argued that such oscillations were local in scale and effect, since patently not all rocks show rhythmic oscillations on Milankovitch time-scales (although it should be noted that many of the examples above relate to sea-level fluctuations, which must be global). However, the Earth's orbit is subject to variation with time, which must be a permanent feature, and which does affect global climates (see Chapter 3). Even with no observed sedimentary oscillations, it would still be expected that there had been global climatic fluctuations at the 20–100-kyr scale throughout Earth history. It turns out that global climatic fluctuations at this frequency have not only occurred but have left a sufficiently strong sedimentary record that they are being proposed as fundamental geological units.

5 · *Biological response: distribution*

How do organisms respond to orbitally-forced climate change? Chapter 4 demonstrated that climates have been changing globally with frequencies on Milankovitch time-scales (10–100 kyr) throughout Earth history. However, it is only during the most recent part of the geological record that sediments have been sampled finely enough, and there is a time-scale precise enough, to make possible useful statements about the response of organisms to climatic change. The climatic changes of the Quaternary were, of course, exaggerated locally by the development of continental ice-sheets with consequent impacts on neighbouring regions. Nevertheless, many parts of the world experienced smaller degrees of change that have been similar to the magnitude of changes experienced during parts of Earth history when continental glaciation was minimal or absent. The modelling experiment of Glancy *et al.* (1986) suggests that Cretaceous geography (apparently with no glaciation (Hambrey & Harland 1981)), at least, could produce temperature and precipitation changes as severe as Quaternary changes in response to varying orbital parameters. There is, in any case, little choice; it so happens that the portion of the geological record most accessible to us (the Quaternary) was a period of fluctuating glaciation, and the response of organisms to climatic change cannot be studied with anything like the same detail in other parts of the geological record. Even within the Quaternary, the evidence is, unfortunately, patchy, with particularly awkward gaps in tropical regions, which may be the best modern parallel for conditions in much of the pre-Quaternary record. In part, these gaps are due to a lack of suitable sediments and in part they are due to a concentration of fieldwork effort in developed countries, which tend to occur away from low latitudes.

The physical background
Major continental glaciers advanced and retreated many times during the late Cenozoic, probably beginning with the development of continental

The physical background · 93

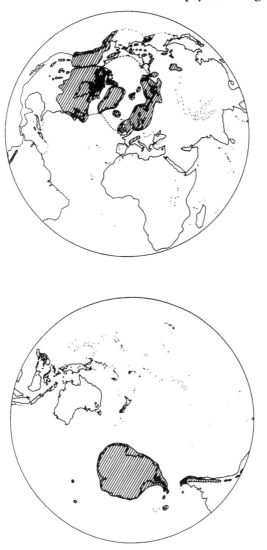

Figure 5.1 Distribution of ice-sheets (shaded) at the time of the last glacial maximum (18 ka). Redrawn by S.M. Peglar from Denton & Hughes (1981).

ice on Antarctica during the late Oligocene–early Miocene, and mountain glaciers elsewhere from the middle Miocene and Pliocene onwards (Hambrey & Harland 1981). The extent of ice-sheets during the last glaciation, at its maximum extent at about 18 ka (Fig. 5.1), may be taken as a typical configuration, but differing in detail from the extent of ice-sheets in earlier glacial periods. The extension of ice-sheets produced a *tabula rasa* over the ground they occupied, bordered by considerable additional extents of land under the influence of a range of periglacial processes. Vast areas, especially in northern North America and northern Europe, have been repeatedly glaciated, and repeatedly re-occupied by plants and animals in the intervening periods.

The other major effect of the build-up of continental ice-sheets was the consequential fall in sea-level as oceanic water became locked up on land as ice, followed by rise of sea-level as the ice-sheet melted. Sea-levels during the Quaternary have fluctuated by as much as 100 m, making substantial changes in the relative areas of land and sea. Modern shallow seas have been repeatedly exposed as dry land during the Quaternary, including the North Sea (northwest Europe), the Java Sea (southeast Asia), and the Bering Strait (between North America and northeast Asia). In these areas, fluctuations in sea-levels have had profound influences on distributions of organisms by permitting or blocking possibilities for colonization. Sea-levels are one aspect of Quaternary environmental changes that have had similar amplitudes right across the globe, regardless of local climate. The effects on coastal organisms must have been substantial on all types of coasts and coastal topographic gradients, and including tropical as well as temperate and polar regions.

An animated sequence of the main changes in global physical environment over the glacial–Holocene transition has been made available by Peltier (1995).

The terrestrial record

The record of climatic oscillations on land is usually, and conveniently, discussed in terms of 'glacial' and 'interglacial' phases, where one pair of phases corresponds to one complete climatic oscillation, as seen in, for example, the $\delta^{18}O$ record (see Fig. 4.3). Minor fluctuations, visible in the pollen record as changes in frequency of forest tree populations, are often referred to as 'stadial' and 'interstadial' (cool and warm, respectively). More informally, the terms 'cold stage' and 'warm stage' can be used to refer to any phase where the fossil evidence indicates the appropriate temperature shift. This terminology originated in those parts of the northern hemisphere where climatic changes are dominated by glacial advance and retreat, and hence climatic changes tend to be dominated by temperature.

The maps of model predictions of climates at 18 ka and 9 ka constructed by Kutzbach *et al.* (1993) from computer experiments (see Figs 3 and 3) should be referred to in conjunction with the individual fossil records presented below, as indications of the direction and magnitude of probable climatic changes during the Quaternary at each of these sites.

Plants

There is an excellent record of Quaternary changes in the distribution and abundance of terrestrial plants through the identification of fossil pollen and plant macrofossils in sequences of peat and lake sediments. The resolution of pollen analysis is such that it is generally not possible to make identifications to species level, but identifications to generic level are usually possible. Within the herb phases, as it happens, most identifications are to family level, and it is therefore not possible to arrive at useful conclusions about community change. Where generic recognition is possible (for example *Artemisia*), there are usually too many possible species, growing in a range of habitats, for further analysis. For the trees, although identification is usually only to generic level, the numbers of possible species are sufficiently small that some useful analyses can be made. Additional methodological and interpretative considerations are described fully by Birks & Birks (1980) and Faegri & Iversen (1989).

The plant record is most detailed with respect to Holocene changes following the retreat and disappearance of the last continental ice-sheets in the northern hemisphere, and has enabled an especially detailed reconstruction of the way in which forest communities have developed. The nature of community change is presented by examining sequences that cover a series of consecutive glacial–interglacial oscillations. Three of these sequences are in southern Europe, giving some indication of regional spatial variation in vegetation change over many oscillations, and the others are from North America, South America, and Australasia, giving a global perspective. Data from other regions with records of vegetation during previous glacial–interglacial oscillations that are discontinuous, both spatially and temporally, support the interpretation from these long, continuous records (see, especially, West's (1980) synthesis of work in eastern England).

Europe

Long pollen records The crater lake basin of Valle di Castiglione, near Rome, Italy, has infilled with a sequence of polliniferous lacustrine sediments, from which Follieri et al. (1988) have obtained an 88.25-m core. Follieri et al. (1988) and Magri (1989) constructed a chronology for the sequence by establishing rates of sediment accumulation for the Holocene using radiocarbon age determinations (Alessio et al. 1986), and by obtaining rates for a lower section from counts of couplets of laminations that had been shown to be annual. Both methods yielded values of 0.31 mm yr^{-1}. They then showed that there was a clear correspondence between phases of high tree pollen concentrations and the timing of precessional phases, as calculated by Berger (1978a), if a uniform accumulation rate of 0.32 mm yr^{-1} was assumed. On this basis, they infered that the rate of sedimentation was close to linear throughout the sequence, and used this as the basis for a chronology.

The sequence at Valle di Castiglione covers the Quaternary since 270 ka, with about 18 oscillations corresponding, broadly, to a 'warm stage' and adjacent 'cold stage' of one oscillation of climatic change. The pollen record (Fig. 5.2) is dominated by phases of herb pollen (*Artemisia*, Chenopodiaceae, and Gramineae) alternating with phases of tree pollen types, interpreted as a series of periods of forest growth around the site, separated by periods of steppe vegetation, dominated by herbs. These latter are contemporaneous with glacial periods further north in Europe (and elsewhere). There are long-term changes, such as the loss in the later part of the sequence of the trees *Pterocarya* (not included on Fig. 5.2) and *Zelkova*, which appear to have become regionally extinct at about 190 ka and 40 ka, respectively. The abundances of other tree pollen types rise and fall, so that no two forested phases are identical in terms of the trees present or their relative abundances. Some tree taxa, such as deciduous *Quercus*, tend to be frequent during most forested phases (but not the forest phase at about 130 ka), while other genera reach high frequencies only in one or a few phases, such as *Ulmus* (at about 250 ka and 90 ka), *Alnus* (at about 5 ka), or *Olea* (at about 130 ka). Although phases of forest and non-forest can generally be identified, and correlated with variations in the Earth's orbital parameters, within forested phases there is no sign of any repeating unit of plant community. Forest development in each warm stage followed a unique course, not repeated before or since.

Cores from the fault-bounded basin of Tenaghi Philippon in northeast Greece have been investigated palynologically by Wijmstra (1969), Wijmstra & Smit (1976), and van der Wiel & Wijmstra (1987a, b). The

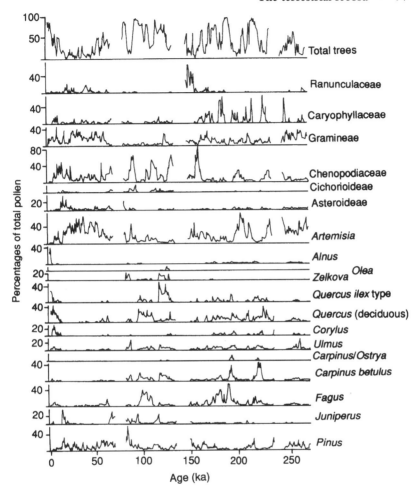

Figure 5.2 Abundances of pollen types in the sequence of Quaternary lake sediments at Valle di Castiglione, Italy. Redrawn from Follieri *et al.* (1988, Figs 3, 5, and 7).

basin covers an area of about 55 km², and is infilled with clay, peat, and marl sediments to a depth of 280 m. The basal 83 m of sediment do not contain pollen, but the rest of the sequence has been analysed from two cores, the first covering the upper 120 m of the sequence, and the second used to cover the sediments at 112.8–197.8 m. The sequence is dated, within broad limits, by radiocarbon age determinations at the top, and by magnetic polarity determinations, including the Brunhes–Matuyama magnetic reversal, taken to have occurred at 780 kyr. Mommersteeg

98 · Biological response: distribution

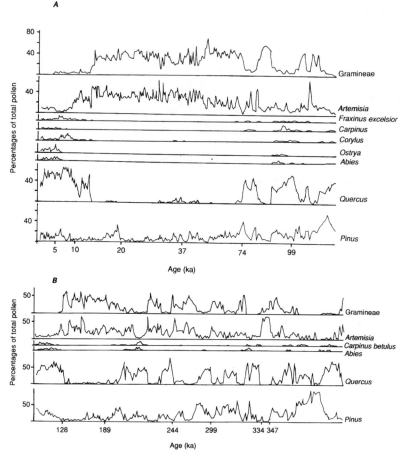

Figure 5.3 Abundances of selected important pollen types in the sequence of Quaternary lake sediments at Tenaghi Philippon, Greece.
A. 0–120 ka (redrawn from Wijmstra 1969, Fig. 2).
B. 120–420 ka (redrawn from Wijmstra & Smit 1976, Fig. 2).
C. 420–650 ka (redrawn from van der Wiel & Wijmstra 1987a, Fig. 2).
D. 650–975 ka (redrawn from van der Wiel & Wijmstra 1987b, Figs 2 and 3).
Ages are those obtained from correlations with oxygen isotope stratigraphy (Mommersteeg *et al.* 1995). The taxon 'Gramineae' (*A* and *B*) is equivalent to 'Poaceae' (*C* and *D*).

et al. (1995) used these ages to derive a time-series for the changing abundance of total tree pollen, and showed that this varies rhythmically with periodicities within Milankovitch time-scales. The timing of peaks in the oscillations closely matches the timing of oxygen isotope stages in

The terrestrial record · 99

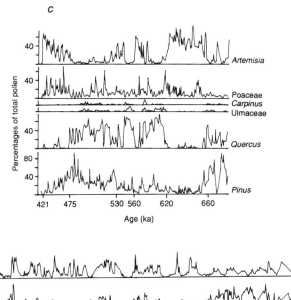

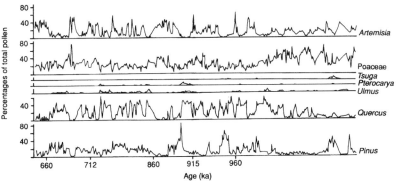

the deep-sea record of Shackleton & Opdyke (1976). The age of the base of the sequence is estimated to be 975 ka.

The pollen stratigraphy (Fig. 5.3) is marked by oscillations between pollen spectra interpreted as 'steppe', 'forest–steppe', and 'forest'. Forest vegetation dominated by a *Quercus* (deciduous) type developed and diminished about 20 times ('interglacials') during the period covered by the sequence. At other times, *Pinus*-dominated forest appeared, but vegetation for most of the intervening periods seems to have been dominated by taxa such as *Artemisia*, Gramineae, and Chenopodiaceae, suggesting steppe-like vegetation. Detailed electron microscope work by Smit & Wijmstra (1970) has suggested that the Chenopodiaceae pollen includes the modern central Asian genus *Krascheninnikovia*, and probably also the genus *Kochia*, indicating not only extreme steppe-like conditions, but also that there is considerable mobility of herbaceous taxa on Quater-

100 · Biological response: distribution

nary time-scales, despite normal concealment of this mobility by the poor taxonomic resolution for most herbaceous pollen types. Forest composition shows considerable variety from one zone to another, even though the forest record has more *Quercus* and *Pinus* than Valle di Castiglione. Thus, there are high frequencies of *Alnus, Castanea, Larix,* and *Tsuga* at about 980 ka, high frequencies of Ulmaceae, *Abies, Betula,* and *Tsuga* at about 960 ka, and high frequencies for *Abies, Corylus, Zelkova, Carpinus,* and *Fraxinus excelsior* at about 550 ka. The amount by which forest composition changes within a forest phase varies considerably. Some stages show no substantial presence of trees other than *Quercus* and *Pinus*, but other stages display a rich variety of trees, with suggestions of some replacement of populations from different genera taking place within the phase. This temporal diversity may be due to the interaction of biotic factors (such as competition), the length of the phase, determined by the contemporary dominant period of orbital forcing, and the amplitude of various components (temperature, precipitation, etc.) within the climatic response to forcing.

During this long record of vegetation change, several tree genera become extinct, including *Cedrus, Celtis, Carya, Eucommia, Liquidambar, Parrotia, Pterocarya, Tsuga,* and *Zelkova*. Some of these genera are now completely absent from Europe and adjacent regions (*Tsuga, Carya, Eucommia,* and *Liquidambar*), while others have severely restricted modern distributions (*Cedrus, Parrotia, Pterocarya,* and *Zelkova*). Such extinctions, within the time represented by the sequence, limit the size of the available species pool, placing long-term constraints on which genera can be represented during any forested phase following the extinction.

A third long Quaternary sequence in Europe is from a tectonic basin at Ioannina, in northwest Greece. Tzedakis (1993, 1994) has presented pollen analyses from 162.75 m of lake sediment, representing the last 423 kyr (Fig. 5.4). There are first order vegetation oscillations between forest and open vegetation, and second order changes within vegetation of either type. Each of the first order oscillations has its own distinct character, although an underlying pattern is discernible. Tree succession in nine forest periods typically involves an initial increase in *Quercus* and Ulmaceae, followed by *Carpinus betulus,* then *Abies* and, often, *Fagus.* Changes are seen in open vegetation periods from steppe–forest vegetation through grassland to a discontinuous desert–steppe vegetation, dominated by *Artemisia*, Chenopodiaceae, and Gramineae. Ioannina is located near to postulated refugia for forest trees in the Balkans during glacial stages (Huntley & Birks 1983; Bennett *et al.* 1991), and Tzedakis

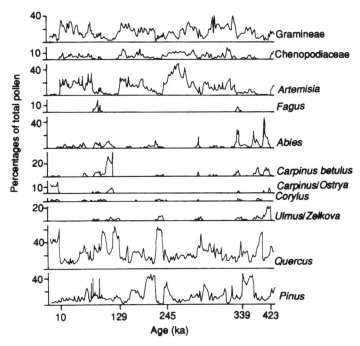

Figure 5.4 Abundances of pollen types in the sequence of Quaternary lake sediments at Ioannina, northwest Greece. Redrawn from Tzedakis (1993, Fig. 3).

(1993) drew attention to the nearly continuous record of pollen of forest trees along the sequence, indicating that these taxa were continuously present in the region, although possibly at low abundances. This in turn indicates that the pollen record is dominated by increase and decrease of populations of forest tree taxa that occur in the southern Balkans throughout glacial–interglacial oscillations, rather than by change in distribution area.

Tzedakis (1994) compared the Ioannina record with those at Tenaghi Philippon and Valle di Castiglione and showed that the last interglacial (about 125 ka) in southern Europe had certain characteristic features in terms of the abundance levels of particular forest trees. *Carpinus betulus* seems to have been a characteristic feature of the later stages of the last interglacial, more abundant than in other interglacials. On the other hand, *Fagus* was unusually scarce. Each interglacial has its own distinct climatic regime (Tzedakis & Bennett 1995), and this may be one factor driving the differential pattern of tree abundances that characterizes each interglacial regionally.

These sites are the only three long sequences currently available in Europe covering multiple climatic oscillations on Milankovitch time-scales. Several other sequences cover the period since the beginning of the last interglacial, thus spanning one complete oscillation, and some useful comparisons can be made (Follieri *et al.* 1988). Valle di Castiglione has a richer tree flora than Tenaghi Philippon throughout. *Fagus*, in particular, was prominent as a forest component at the Italian site, but not in northeast Greece, through the period during which the cores overlap. *Zelkova* was prominent at Valle di Castiglione during zone VdC-10 (the last interglacial, or Eemian), but is otherwise unknown from continental Europe in the late-Quaternary (Follieri *et al.* 1986). Expansion of *Fagus* populations occurred at Valle di Castiglione in VdC-12, interpreted as an interstadial during the early part of the last glacial phase, and also at Les Echets, central France, zone D (de Beaulieu & Reille 1984), but not at other sites of similar age. *Olea* was also prominent at Valle di Castiglione and Ioannina (during the last interglacial), but was not abundant elsewhere in Europe. The last interglacial pollen records at Samerberg, southern Germany (Pröbstl & Grüger 1979), and Lac du Bouchet, eastern France (Reille & de Beaulieu 1988) are dominated by *Picea*, which was also an important forest component during the interstadials of the early part of the last glacial at Grande Pile, northern France (Woillard 1978; de Beaulieu & Reille 1992) and Les Echets, but has only a sporadic presence at Valle di Castiglione and Tenaghi Philippon. In contrast to these differences, *Carpinus betulus* is consistently present and abundant at all European sites shown to be Eemian in continuous sediment sequences (Follieri *et al.* 1988).

These records display considerable variation across continental Europe in the nature of different forest stages. In any one sequence the composition of forest differs from one stage to another, and the specific representation in forests of one stage also varies spatially. However, the uniform abundance of *Carpinus betulus* across Europe during the Eemian has not been repeated during the Holocene for *Carpinus betulus* or for any other tree species.

Holocene pollen records   There are now hundreds of Holocene pollen sequences available in Europe (Huntley & Birks 1983; Huntley & Prentice 1993; Peterson 1993; Willis 1994), mostly concentrated in the formerly glaciated areas of northern Europe because of the widespread occurrence there of sediment sequences in lake basins created by glacial action. Synthetic work on these records has begun (Huntley & Birks 1983; Peterson

1993), and is yielding information on the assembly of plant communities that is complementary to the record from the much smaller number of long sequences. Broadly, the pattern is one of tree populations establishing and spreading the overall distribution of species (and hence genera) out of glacial refugia and into regions made accessible by increasing warmth and glacial retreat.

Huntley & Birks (1983) presented a series of maps that established the changing distribution of pollen types in space and time across the whole of Europe (Figs 5.5 and 5.6). Their maps were compiled by collecting data on the frequency of each pollen type at each site for 500-yr intervals of time during the Holocene and late last-glacial (since 13 ka). These maps show, principally for forest trees, that each type has spread independently, at different rates, in different directions, and at different times. Thus, *Quercus* spread early and fast from south to north (Fig. 5.5), while *Tilia* spread later and more slowly from east to west (Fig. 5.6). The compositions of modern forests have no long histories, but are the outcomes of the sum of responses of each tree type following the climatic warming at the end of the last glacial.

Using the same dataset, Huntley (1988, 1990b), treating each pollen spectrum from each site at each time slice as an individual data item, carried out a multivariate classification of the data and plotted the resulting clusters in time and space, thus displaying the distribution of pollen assemblages in time and space (Fig. 5.7). The maps shown indicate the temporal and spatial extent of plant communities, strengthening earlier conclusions (for example Iversen 1960; West 1964; Huntley & Birks 1983) that plant communities are impermanent (1–10 kyr) assemblages resulting from the individualistic behaviour of taxa in response to environmental changes. As a corollary, they expose the lack of modern analogues for many of the pollen assemblages seen in the fossil record. This last point is nicely illustrated by dissimilarity mapping between past pollen assemblages and modern pollen assemblages to discover, quantitatively, the difference between modern and fossil communities and the way in which that difference has changed in moving towards the present (Fig. 5.8). Vegetation units for which there is no modern analogue have existed at all times since 13 ka, to as recently as 1 ka in some areas (Huntley 1990a).

At the smaller scale of the British Isles, Birks (1989) mapped the changing distribution limits of the principal forest trees during the Holocene. The areal extent of each genus has increased through time independently of the others (Fig. 5.9). Late arrivals tend to occupy ground more

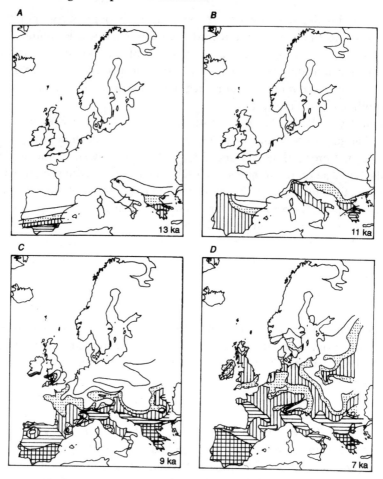

Figure 5.5 Maps of changing distribution and abundance of deciduous *Quercus* (oak) in the late-Quaternary of Europe. Between-contour shading indicates areas where, at the times indicated, sites have *Quercus* (deciduous) pollen frequencies of 2–5% (unshaded), 5–10% (stipple), 10–25% (vertical shading), 25–50% (horizontal shading), and > 50% (cross-hatch). Frequencies are percentages of total pollen (*A* and *B*) or of total tree and shrub pollen (*C* and *D*). These different calculation bases yield similar results within the time periods shown here. Redrawn from Huntley & Birks (1983, Figs 5.221, 5.223, 5.227, and 5.229).

slowly, but *Corylus* spread more rapidly than *Betula*, which preceded it. The development of forests must have shown considerable regional variation depending on the local time of appearance of the various modern constituents of the forest. Species combinations have been in contact

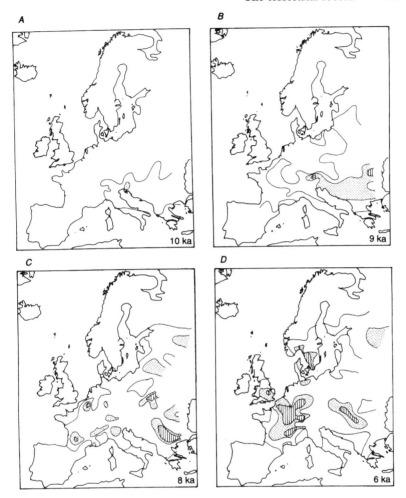

Figure 5.6 A–D Maps of changing distribution and abundance of *Tilia* (lime) in the late-Quaternary of Europe. Between-contour shading indicates areas where, at the times indicated, sites have *Tilia* pollen frequencies of 1–5% (unshaded), 5–10% (stipple), and 10–25% (vertical shading). Frequencies are percentages of total tree and shrub pollen. Redrawn from Huntley & Birks (1983, Figs 5.255, 5.257, 5.259, and 5.261).

with each other for differing periods of time in different parts of the islands. Some of this variation is due to varying rates of spread from the regions, mainly in southeastern Europe, where populations survived the last glacial period (Huntley & Birks 1983; Bennett *et al.* 1991). How-

106 · Biological response: distribution

A

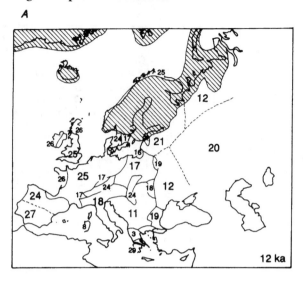

B

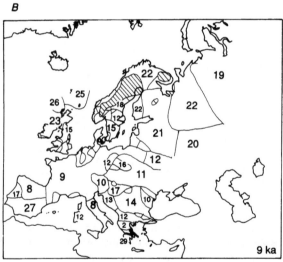

Figure 5.7 A–D 'Vegetation' units identified by two-way indicator species analysis (TWINSPAN) of late-Quaternary pollen data from Europe. The numbers identify areas as having pollen samples corresponding to one of 29 'vegetation' units. Ice-sheets are shown as shaded areas. Redrawn from Huntley (1988, Figs 6, 8, 11, and 12).

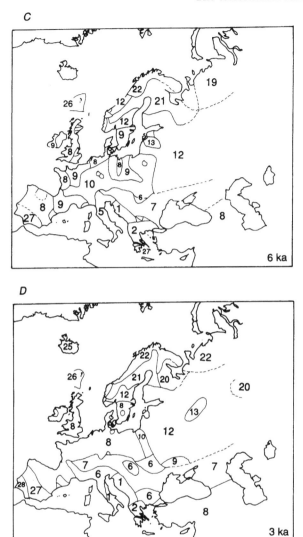

Figure 5.7 (continued) For legend see page 106.

ever, some of it is due to slow rates of increase of populations of species such as *Alnus glutinosa*, which have localized habitats for growth and demanding requirements for successful reproduction that combine to make the occupation of suitable sites a slow and haphazard process (Bennett & Birks 1990).

108 · Biological response: distribution

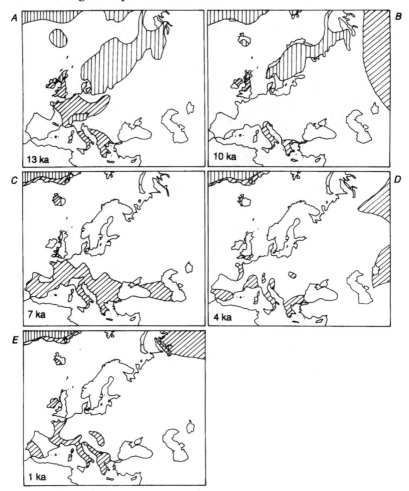

Figure 5.8 A–E Time-series of areas of no-analogue vegetation in Europe since 13 ka. Diagonal shading indicates regions where minimum chord distance values between fossil pollen assemblages and modern pollen assemblages exceed 0.3, indicating that there are no modern pollen samples that resemble the fossil samples. Vertical shading indicates extent of ice-sheets. Redrawn from Huntley (1990a, Fig. 3).

North America

Botanical evidence from North America for Quaternary vegetation and floral changes comes from both the fossil pollen approach, and the more recently discovered macrofossil record of pack-rat middens (Wells & Jorgensen, 1964). The two types of data are complementary, each contri-

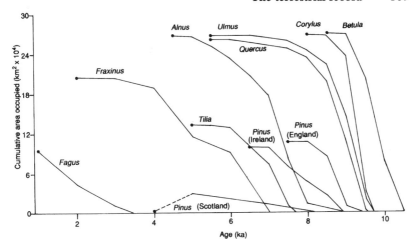

Figure 5.9 Cumulative area occupied in the British Isles by different forest trees during the Holocene. Redrawn from Birks (1989, Fig. 13).

buting to the whole problem from different aspects in a way that has not yet been achieved on any other continent.

Long pollen records   To date only one long sequence of sediment, covering at least the period since the beginning of the last interglacial, has been recovered and analysed. This is from Clear Lake, an extant lake in a structural basin of complex origin in the northern Coast Ranges of California, from which a 115-m sediment core has been studied palynologically and described by Adam (1988). Present vegetation around the lake consists mostly of woodland dominated by populations of the trees *Quercus douglasii* and *Pinus sabiniana* with some chapparal, dominated by evergreen sclerophyll shrubs (*Adenostoma*, *Arctostaphylos* and *Ceanothus*), and mixed hardwood forest dominated by populations of *Arbutus menziesii*, *Pseudotsuga menziesii* and from species in several genera of the Fagaceae. Coniferous forest occurs at higher elevations in the mountains. The lake sediments consist uniformly of clays, but with organic content increasing gradually upwards. The age of the core has been estimated from radiocarbon age determinations of the upper sediments and extrapolation of the calculated accumulation rate to the base of the sequence. This gives a basal age of about 130 ka, which turns out to be consistent with the time-scale from the deep-sea $\delta^{18}O$ record, assuming that changes in the *Quercus* pollen record can be correlated with $\delta^{18}O$ (S.W. Robinson, cited in Adam 1988).

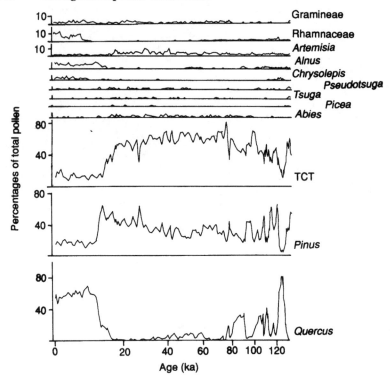

Figure 5.10 Abundances of pollen types in the Quaternary lake sediments at Clear Lake, California, USA. TCT, Taxodiaceae, Cupressaceae, and Taxaceae. Redrawn from Adam (1988, Plate 1).

The pollen record from Clear Lake (Fig. 5.10) is dominated numerically by pollen of the trees *Quercus*, *Pinus* and a category comprising the Taxodiaceae, Cupressaceae, and Taxaceae (TCT). Adam (1988) divided it into four main climatic units: a basal cold stage (Tsabel: before 128 ka), succeeded by a warm stage (Konocti: 128–120 ka) correlated with the last interglacial in North America and elsewhere, then a cold stage (Pomo: 120–10 ka), which is subdivided into a number of minor warm and cold stages, and ending with a warm stage (Tuleyome) that covers the last 10 kyr to the present. The Konocti warm stage, which is the only unit inferred to have had a temperature at least as warm as the present, has pollen spectra dominated by *Quercus*, with higher values than in the Holocene, and negligible representation of other tree types. The Tuleyome warm stage (the Holocene) is also dominated by *Quercus*, but has significant amounts of pollen from *Chrysolepis* and *Alnus*, and a much greater abundance of the Rhamnaceae pollen, which would include that from the chapparal genera *Ceanothus* and *Rhamnus*. The cold stages are dominated by populations of *Pinus* and TCT, with some *Abies*, *Picea*, *Tsuga* and *Pseudotsuga*, in varying amounts. *Tsuga* and *Pseudotsuga* tend to be more abundant in the Tsabel cold stage, while *Abies* and *Picea* tend to be more abundant in the Pomo cold stage.

Broadly, this record can be considered as representing elevational belts moving up and down mountains in response to climatic warming and cooling. But in so doing, it is clear that the proportions of various populations in the belts changed, some being more abundant in some stages than others. It also appears that chapparal may be a Holocene novelty, at least in its present areal extent.

Holocene pollen records The North American Holocene pollen record has been well-studied, especially in the northern and eastern part of the continent. Davis (1976, 1981b, 1984) has produced maps of changing range limits of the main forest trees of the east, plotting contours ('isochrones') on the time of increase of each pollen type at each of the sites in the area (Fig. 5.11). Additionally, maps illustrating the pollen abundance of each of 12 taxa of eastern North American forest trees at 250-yr intervals from 15 ka to the present have been made available by Keltner (1995). It is apparent from these maps that the various forest types of eastern North America have different histories, as their constituents spread at different rates and in different directions. Even forests belonging to the same type today may have had different histories. The *Quercus–Castanea* forests of the central Appalachians, for example, have been in existence for 5 kyr or more, but the *Quercus–Castanea* forests of Connecticut have only developed since the arrival of *Castanea* 2000 years ago. Ohio deciduous forests included *Carya* 4 kyr before *Fagus* arrived, but in Connecticut *Fagus* preceded *Carya* by 3 kyr. Four tree taxa that now have overlapping, nearly coincident ranges in the modern boreal forest (*Picea*, *Larix laricina*, *Abies*, and *Pinus banksiana/resinosa*) had different distributions earlier in the Holocene. Davis (1981b) argued that the climatic changes at the beginning of the Holocene were the ultimate cause of these tree 'migrations', but that spread of each type was then limited by factors such as seed dispersal and competition. She concluded by pointing out that those studying 'adaptations of species' that allow them 'to partition the environment and avoid competition' should examine the fossil record and ascertain whether populations from the species of

112 · Biological response: distribution

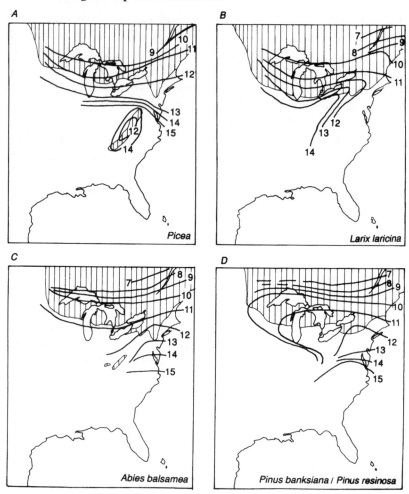

Figure 5.11 A–L. Maps of the late-Quaternary spread of tree pollen types in eastern North America. Shaded area indicates modern distribution. Numbered lines refer to the time (ka) since the first increase of the type after 15 ka, representing the leading edge of the spreading population. Redrawn from Davis (1981b, Figs 10.3–10.14).

interest have co-occurred long enough for such adaptations to develop (Davis 1976, 1981b).

Maps illustrating the spread of a variety of trees and herbs, as inferred from pollen data, have also been presented by Webb (1987, 1988). In contrast to Davis (1976, 1981b, 1984; Davis *et al.* 1986), he and

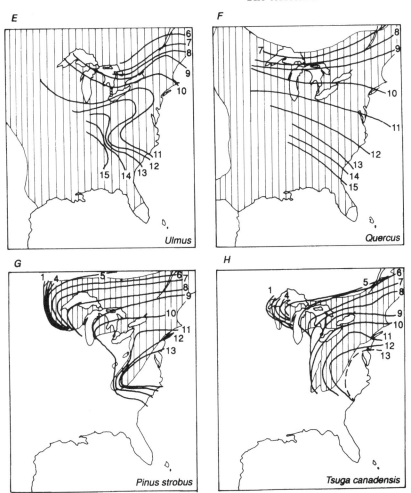

Figure 5.11 (continued) For legend see page 112.

his colleagues have argued that climatic change within the Holocene has been the main controlling factor of Holocene distribution changes (Webb 1986; Prentice *et al.* 1991: see also page 41). However, there is consensus that the trees have spread individualistically across the eastern continent during the early and mid-Holocene, and that in so doing plant communities have formed and some have broken up so that there has not necessarily been any long contact between any pair or group of taxa. Jacobson *et al.* (1987) combined the distributions of various taxa

114 · Biological response: distribution

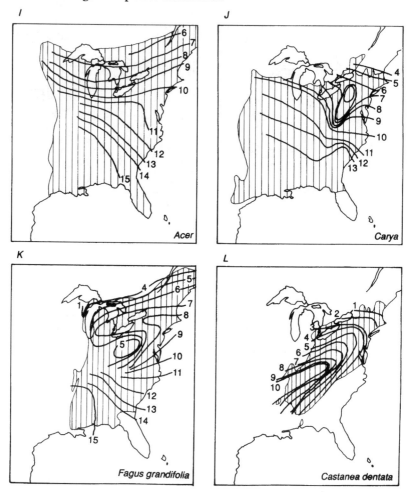

Figure 5.11 (continued) For legend see page 112.

that comprise modern vegetation units to illustrate (by the extent of the area of overlap in the past) how that unit has changed. Thus, the distributions of the trees *Tsuga* and *Fagus* were distinct at 12 ka and 10 ka, overlapped slightly in the Massachusetts area at 8 ka, and reached their present substantially coincident ranges by 6 ka.

Davis (1976, 1981b) suggested that the main forest trees of eastern North America had extended their ranges at rates of 100–400 m yr^{-1}, and Delcourt & Delcourt (1987) showed how the rate of range extension varied in time and space, taking place more rapidly from some populations of a species' distribution than others. There are problems with the way these rates of spread are calculated, because of possible confusion about how it is possible, palynologically, to distinguish invasion by a population of a species new to the area, and population increase from levels too low to have been detected (Bennett 1986b), but in some cases rapid spread can be demonstrated unequivocally. Ritchie & MacDonald (1986) have demonstrated that *Picea glauca*, one of the dominants of the North American boreal forest, spread at rates of up to 2000 m yr^{-1} towards northwest Canada, along the margin of the retreating Laurentide ice-sheet during the early Holocene. This high rate was probably enabled by the transport of seeds by surface winds moving clockwise along the front of the remnant ice-sheet, as indicated by atmospheric model experiments (Kutzbach & Guetter 1986).

Ritchie (1987a), in particular, has pointed out that claims about past 'no-analogue' vegetation need to be backed up by quantitative analyses, rather than the more typical descriptive accounts. Anderson *et al.* (1989) carried out a comprehensive analysis of pollen spectra from 12 sites in western Canada and interior Alaska. Plots of the dissimilarity between each pollen spectrum and the modern sample that it most closely resembles (Fig. 5.12) show changing patterns of approach between the fossil pollen spectra and modern vegetation. At some sites, modern conditions are approached gradually over many millenia (for example Kaiyak Lake), but at others there is an abrupt transition (for example Tiinkdhul Lake, at 7.5 ka). Anderson *et al.* (1989) showed that full-glacial vegetation is not similar to any modern vegetation, but most resembles modern tundra. Boreal forests originated only during the Holocene. Similarly, Overpeck *et al.* (1992, 1995) found that vegetation types with no modern analogue were widespread in eastern North America before about 10 ka.

Packrat middens   The macroscopic remains of plants accumulated by packrats and woodrats (*Neotoma* spp.) and preserved in the crystallized urine of their middens constitute a remarkable source of palaeoecological information for the period of the last glacial to the present (Betancourt *et al.* 1990). Each midden can be dated using the radiocarbon method, giving an age for a set of macrofossils, often identified at species level, at a particular place. Many dated middens over a region can thus provide evidence of past distributions of plants in space and time. Preserved packrat middens are found mainly in the arid southwest of the USA and northern Mexico, which are areas where there are few or no suitable sites

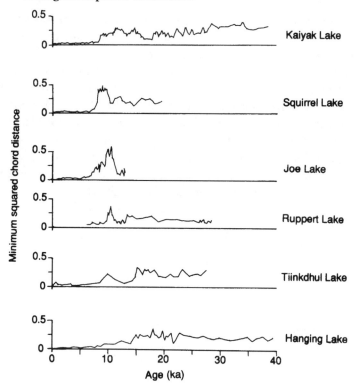

Figure 5.12 Dissimilarity values (minimum squared chord distance) between each fossil spectrum in six late-Quaternary pollen sequences and the modern sample it most closely resembles. The six sites are from northwestern Canada and adjacent Alaska. Redrawn from Anderson *et al.* (1989, Fig. 9).

for preservation of pollen-bearing sediments. Packrat-midden records lack the continuity of pollen records from lake sediments, but gain in their level of taxonomic resolution. Methodological and interpretative aspects of this research are discussed by Dial & Czaplewski (1990), Finley (1990), Spaulding *et al.* (1990), and Webb & Betancourt (1990).

Regional syntheses have been recently published for most areas of the deserts of Mexico and the southwestern USA. In the Chihuahuan Desert, community composition has changed continuously over the last 11,000 years. At the end of the last glacial pinyon (*Pinus remota*), juniper (*Juniperus* spp.), and other plants retreated from desert lowlands as woodland, grassland, and desert scrub taxa spread to new areas. Present Chihuahuan desert communities were only completed in the last 4–5 kyr (van Devender 1990a). Van Devender (1990b) listed 46 Sonoran Desert species pairs found together in middens of the last glacial, or Holocene, and not associated today. These include *Artemisia tridentata* from modern Great Basin desert scrub and *Echinocactus horizonthalonius* from modern Sonoran desert scrub; *Vauquelinia californica* from modern desert grassland and *Cupressus arizonica* from modern woodland; *Symphoricarpos* sp. from modern woodland and *Castela emoryi* from modern Sonoran desert scrub. Modern plant communities did not develop until the late Holocene.

Spaulding (1990) recorded complex patterns as Mojave Desert communities changed composition between the last glacial and the present. Some last glacial plants became extinct, and there was a group of early Holocene immigrants that subsequently disappeared. Holocene immigration was staggered. At one site *Larrea divaricata* and *Ambrosia dumosa* occurred together at 7.9 ka, but further west, where *Ambrosia dumosa* was present by 7.9 ka, *Larrea divaricata* did not arrive until 5.9 ka. The two plants are common associates today, but have had different distribution histories in the past. A summary of present and last glacial altitudinal ranges of selected trees and shrubs (Fig. 5.13) shows that most now have higher distributions than during the last glacial, but some have barely changed (for example *Fallugia paradoxa*), and one (*Atriplex confertifolia*) now occurs at lower altitude than during the last glacial (Spaulding *et al.* 1983). Several species of tree appear to have spread through the area during a brief interval of nearly continuous woodland and established populations on mountain tops that were inhospitable during the last glacial, and hospitable but inaccessible because of the intervening desert for much of the Holocene.

Thompson (1990) found that the last-glacial flora of the Great Basin was impoverished relative to today. Holocene communities developed by dispersal of populations from new species to the area, and extinction of the last populations from some last-glacial species. This development took place in different fashions in different sites. *Ephedra viridis* was present along the southern margins of the Basin by the beginning of the Holocene, but did not reach the central part until 4 ka. In eastern Nevada, a woodland community including *Pinus flexilis, Juniperus scopulorum, Pinus monophylla,* and *Juniperus osteosperma* existed between 7 ka and 6 ka, but then the first two species became extinct. Most modern communities were established by 6 ka, but the establishment and extinction of some species was still sufficiently dynamic that some overall species distributions were still changing in the late Holocene. The complex mountain topography of the area means that the spread of woodland and

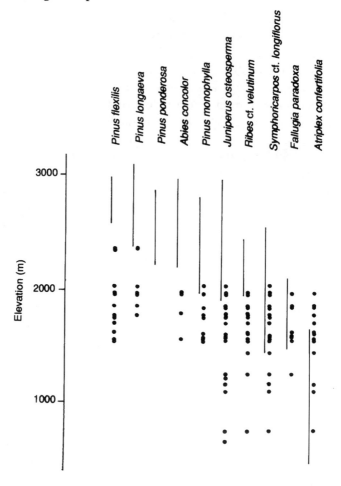

Figure 5.13 The modern and full-glacial ranges of selected trees and shrubs on calcareous substrate in southern Nevada, USA. Modern elevational range is indicated by the continuous lines, and the full-glacial range by filled circles. All full-glacial records are from packrat (*Neotoma*) middens older than 11 ka. Redrawn from Spaulding *et al.* (1983, Fig. 14.7).

montane plants became more difficult later in the Holocene as altitudinal woodland moved up the mountains, and these habitats became more restricted. Fig. 5.14 illustrates the establishment of modern communities in this area.

The development of Holocene communities in the Grand Canyon area is discussed by Cole (1985, 1990). His results (Fig. 5.15) display

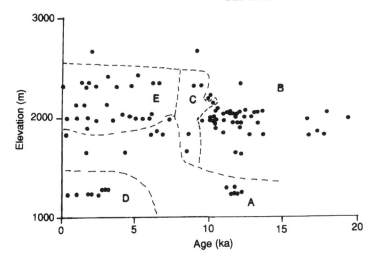

Figure 5.14 Zonation of vegetational assemblages in the Great Basin, USA, since 20 ka. Data points are radiocarbon-dated packrat (*Neotoma*) middens. Assemblages are: A, *Juniperus*; B, subalpine; C, montane; D, desert; E, *Pinus–Juniperus*. Redrawn from Thompson (1990, Fig. 10.12).

elegantly a period of community disruption and reformation between about 12 ka and 6 ka, involving a complete re-sorting of populations of the available species, plus some new species and minus a few that had become regionally extinct, into new communities with no history in the area longer than the mid-Holocene. Early Holocene middens contain plant assemblages that are quite unlike modern or late-glacial assemblages (Cole 1990). Cole (1985) used Sørenson's index of similarity to evaluate the difference between plant associations found in fossil middens and those in modern middens (Fig. 5.16). This depicts, in a different way, the late development of communities that had anything like the modern character.

Fig. 5.17 illustrates the changes in plant altitudinal zonation on the Colorado plateau between the late-glacial and present (Betancourt 1990). Although many of the same species are involved, the zonation has changed completely. Some species, dominant in the last glacial, have lost their status (for example *Juniperus scopulorum*), while others that were completely absent during the last glacial have become dominant in the Holocene (notably *Pinus flexilis* and *Pinus ponderosa*). Species richness appears to have been lowest at the climatic transition of the beginning of the Holocene, probably because extinctions preceded arrivals (as in

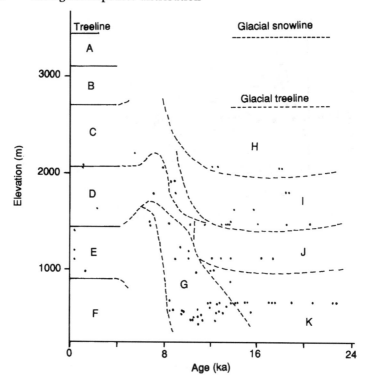

Figure 5.15 Zonation of major plant communities in northern Arizona, USA, since 24 ka. Data points are radiocarbon-dated packrat (*Neotoma*) middens. Middens below 800 m elevation are from the western Grand Canyon, and the others are from the eastern Grand Canyon. Full-glacial treeline and snowline elevations are inferred from pollen data and glacial deposits, respectively. Community dominants are: A, alpine conifers; B, *Picea*; C, *Abies concolor*, *Pseudotsuga menziesii*, and *Pinus ponderosa*; D, *Pinus* and *Juniperus*; E, *Coleogyne ramosissima* and *Artemisia* sec. tridentatae; F, *Encelia farinosa* and *Larrea divaricata*; G, *Juniperus* and *Fraxinus anomala*; H, *Picea* and *Juniperus*; I, *Pinus flexilis*, *Abies concolor*, and *Pseudotsuga menziesii*; J, *Juniperus*, *Atriplex confertifolia*, and *Artemisia* sec. tridentatae; K, *Juniperus*, *Coleogyne ramosissima*, and *Atriplex confertifolia*. Redrawn from Cole (1985, Fig. 1).

the Grand Canyon: Cole (1985, 1990)). There appear to have been some species that expanded their ranges during the early Holocene, then contracted later (for example *Pinus ponderosa* and *Quercus gambelii*), while many others only achieved maximum ranges in the late Holocene.

The packrat-midden results described above are important contributions to understanding of late-Quaternary palaeoecology (Ritchie 1995).

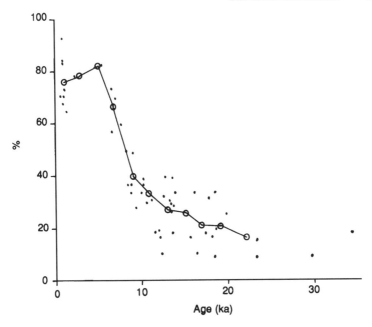

Figure 5.16 Sørenson's index of similarity comparing species in Grand Canyon packrat (*Neotoma*) midden fossil deposits with species present within a radius of 30 m of the midden today. Data points give the value for the index against the radiocarbon age of the midden. Open circles are 2000 yr average values for the index, connected by straight lines to indicate trend. Sørenson's index is calculated as $2C/(A + B)$, where A is the number of species at the site today, B is the number of species in the midden, and C is the number of species in common. Redrawn from Cole (1985, Fig. 2).

Most of our knowledge of how plants respond to Quaternary climatic change comes from pollen with a poorer taxonomic resolution. The higher resolution of the fossils from middens shows that the type of plant behaviour inferred from the generic or familial level of pollen taxonomy can also be seen at the specific level, which is reassuring, but it is also an important contribution to evidence for the individualistic behaviour of populations in response to Quaternary climatic changes.

Most packrat-midden work has focussed on trees and shrubs, but recent efforts with more difficult groups, such as grasses (van Devender *et al.* 1990) show similar patterns, and hold promise for future results based on a wider range of taxonomic groups and life forms than is currently possible with any fossil material.

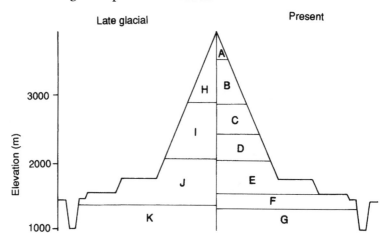

Figure 5.17 Generalized late glacial and present plant zonation on the Colorado Plateau, USA. Vegetation zones are: A, alpine; B, *Picea engelmannii, Abies lasiocarpa*; C, *Pseudotsuga, Abies concolor, Picea pungens*; D, *Pinus ponderosa*; E, *Pinus edulis – Juniperus osteosperma/monosperma* woodland; F, *Juniperus osteosperma/monosperma* grassland; G, Desertscrub – grassland; H, alpine – *Artemisia*; I, *Picea engelmannii, Abies lasiocarpa*; J, *Pinus flexilis, Pseudotsuga, Picea pungens, Abies concolor, Juniperus scopulorum, Artemisia* steppe; K, *Juniperus osteosperma, Artemisia* halophytic desertscrub. Redrawn from Betancourt (1990, Fig. 12.11).

Central America

Lacustrine deposits in an extinct caldera at El Valle, Panama (500 m above sea level), have yielded a 55-m sediment core (Bush & Colinvaux 1990). Radiocarbon age determinations demonstrate that the record extends back at least 150 kyr, thus spanning the complete oscillation from last interglacial to the present interglacial. Bush & Colinvaux (1990) interpreted the pollen record (summarized in Fig. 5.18) in terms of the descent of montane forest, including *Quercus*, Sapotaceae, Chenopodiaceae, *Cyathea*, and *Alternanthera*, to lower elevations. Pollen taxa of lowland vegetation include *Byrsonima*, *Cecropia*, Myrtaceae, Melastomataceae, and *Hura*. The last interglacial part of the sequence has high values for the lowland types, while the full-glacial section has a sharp increase in the abundance of montane types. Bush & Colinvaux (1990) noted, however, that lowland forest taxa persisted around El Valle throughout the glacial period, despite the increase in abundance of the montane taxa, creating communities for which there is no modern analogue. They also pointed out that montane floras of east and west Panama cannot have

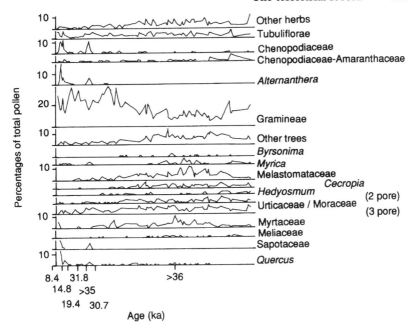

Figure 5.18 Abundances of pollen types in late-Quaternary lake sediments from El Valle, Panama. Redrawn from Bush & Colinvaux (1990, Fig. 2).

merged during the last glacial because the montane flora did not extend low enough, and so disjunctions in the modern distributions of montane taxa such as *Quercus* must have arisen through dispersal rather than fragmentation of more extensive cold-stage distributions. The highlands of Panama were not refugia for tropical rain-forest taxa during any part of the Quaternary (see also Bush *et al.* 1992).

The sediments of a lake at about 100 m above sea-level on the limestone platform of the southern portion of the Yucatan peninsula provide information about low-altitude, low-latitude vegetational change. The site, Lake Quexil, in the Peten district of Guatemala, has a record spanning the period from 36 ka to the present (Leyden *et al.* 1994). Pollen results (Fig. 5.19) reveal the presence of a sparse temperate thorn scrub around the lake during the time of the last glaciation, when sedimentological evidence indicates that the climate of the area was extremely arid. At the end of the last glacial period, temperate oak forest developed, and then mesic tropical forest, predominantly members of the family Moraceae, from about 8.5 ka. During this time, lake sediments indicate

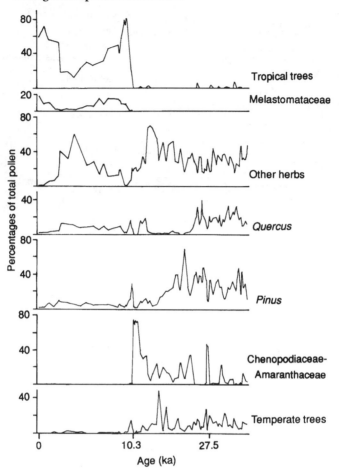

Figure 5.19 Abundances of pollen types in late-Quaternary lake sediments from Lake Quexil, Guatemala. 'Temperate trees' include the mesic and xeric taxa: *Juglans*, *Cornus*, Onagraceae, *Ostrya*, *Carpinus*, *Betula*, *Fraxinus*, *Rapanea*, *Ilex*, *Hedyosmum*, and *Juniperus*. 'Tropical trees' are predominantly Moraceae. Redrawn from Leyden *et al.* (1994, Fig. 2).

that the climate became less arid as well as becoming warmer. Leyden *et al.* (1994) attributed these vegetational changes to climatic changes forced by changing seasonal insolation patterns from orbital variations.

South America

The high plain at Bogotá, Colombia, is an infilled late Cenozoic lake basin, containing at least 800 m of unconsolidated lacustrine clay sedi-

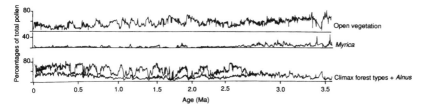

Figure 5.20 Abundances of pollen types in late-Cenozoic lake sediments from the high plain of Bogotá, Colombia. 'Climax forest types' comprise *Quercus*, *Podocarpus*, *Weinmannia*, *Vallea*-type, *Symplocos*, *Drimys*, *Styloceras*, *Daphnopsis*, *Juglans*, Melastomataceae, *Miconia*, Myrtaceae, and *Hedyosmum*. *Alnus* abundances are indicated as the upper curve. 'Open vegetation' comprises *Hypericum* and Gramineae. Redrawn from Hooghiemstra (1984, Diagram 4).

ments. Hooghiemstra (1984, 1989) has presented details of work on a 357-m core (Funza I), within the basin. The results, including pollen analyses, demonstrate that this is 'the longest continuous and detailed climate record available'. It is an unparallelled record of floristic and vegetational change, without equal anywhere in the world. It is impossible to do more than present some outline results and main conclusions from this work.

The sequence includes 110 layers of volcanic ash, from many of which it proved possible to obtain fission-track or K/Ar radiometric ages. This set of age determinations provide the basis for an age model for the sequence. Because the ages are not unequivocal, more than one model is possible, and the original age–depth model has been revised in the light of new determinations from a second core (Funza II) (Hooghiemstra & Sarmiento 1991), but it seems that there has been continuous sedimentation in the basin over the last 3 Ma. Hooghiemstra (1984, 1989) correlated the sequence with the deep-sea $\delta^{18}O$ record, and directly with the calculated record of variation of the Earth's orbital parameters (Berger 1976), and included these as part of one of the age models used to date the pollen record. The discussion below follows the age–depth model of Hooghiemstra & Sarmiento (1991).

The pollen record of Funza I (Fig. 5.20) is subdivided into 55 zones, corresponding to at least 27 climatic oscillations. The spread of oscillation lengths is longer in the part of the sequence after 1.0 Ma (Fig. 5.20). Superimposed on this change of periodicity, the sequence passes through four phases, suggesting some degree of forcing at longer time-scales. The high plain was occupied by predominantly forest vegetation types at 1.4–

0.95 Ma and 0.4–0 Ma, and by open vegetation types in the intervening period (0.95–0.4 Ma).

Apart from the periodicity, the main overall changes in the sequence are the appearance and increase of *Alnus* from about 1 Ma, and *Quercus* (not included separately in Fig. 5.20) from about 340 ka. Both are primarily northern hemisphere tree genera, with pollen records dating from the Oligocene and late Cretaceous (Muller 1981), respectively, so their appearance in South America only at this late date may be a consequence of the tectonic movements that formed the Isthmus of Panama, connecting North and South America in the late Cenozoic. *Alnus* is an abundant pollen producer, and thus has greater prominence in the pollen record than its probable representation on the ground.

The high plain of Bogotá is at an altitude of about 2500 m, within the present-day Andean forest zone, surrounded by mountains rising to above 3600 m. Hooghiemstra (1984) interpreted the pollen record in terms of the altitude of a shifting upper forest limit, at 3250–3500 m today. During the past 3 Myr it has moved between limits of about 3400 m and 1800 m. The major components of the forest were subjected to considerable changes as this took place. Fig. 5.21 illustrates some of the vegetation change in terms of appearances and disappearances of pollen types within the sequence. At any point in the sequence, types come and go, ensuring that, although the vegetation passes through recognizable stages ('Andean forest' and so on), its composition changes from stage to stage to a greater or lesser extent. Additionally, there are changes in the abundances of pollen types between similar stages of the climatic zones.

The combination of long-term changes in flora, such as the appearance of *Alnus* and *Quercus*, periodic changes of vegetation on Milankovitch time-scales, longer periodicities, and taxa present with differing abundances between similar stages of climatic oscillations, all combine to present a highly dynamic view of community change over several million years at this low-latitude site within the South American tropics.

Africa

There are relatively few pollen records from Africa, and none spanning a complete glacial–interglacial oscillation. The most useful is a record of the last 40 ka at Kashiru in the highlands of Burundi at 3°S latitude and 2100 m altitude (Bonnefille & Riollet 1988). The sequence consists of about 10 m of peat that has accumulated continuously at the site. Pollen analyses (Fig. 5.22) show that forest vegetation before 30 ka was dominated by the trees *Podocarpus* and *Macaranga*, the rosaceous shrub

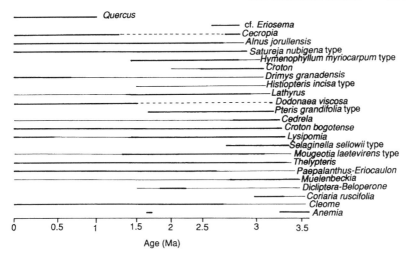

Figure 5.21 Stratigraphic ranges of pollen types in the Quaternary sediment sequence from the high plain of Bogotá, Colombia. Thick lines indicate periods when types are present in almost every sample, thin lines indicate periods of presence in many samples, and dashed lines indicate intermittent presence. Redrawn from Hooghiemstra (1984, Fig. 4).

Cliffortia, members of the Ericaceae, and other genera typical of montane forest. From 30 ka to 10 ka, pollen of Gramineae dominates, with only low proportions of pollen from trees. Pollen of temperate groups, such as Caryophyllaceae, Cruciferae, and *Ranunculus* is also present in quantity. After about 10 ka, pollen frequencies of forest trees increase again to about the same level of abundance as before 30 ka. *Podocarpus*, *Olea*, and *Macaranga* increased first, followed by simultaneous increase of *Alchornea*, *Ficalhoa*, and *Syzygium*, and then by Araliaceae and *Hagenia abyssinica*. Main forest development took place after about 6.7 ka. This sequence illustrates the response of African montane forest vegetation to climatic changes from the last glacial maximum to the present. The magnitude of these climatic changes is unlikely to have been more than a decrease of about 5°C in temperature with or without some precipitation change. Yet, the change of vegetation from forest to herbaceous vegetation and back to forest within 20 kyr can hardly have been more comprehensive. Other pollen sequences from elsewhere in the east African mountains have produced similar results (Flenley 1979).

In the arid centre of the present eastern Sahara desert there are a number of sites where lakes developed during the Holocene, but which have

128 · Biological response: distribution

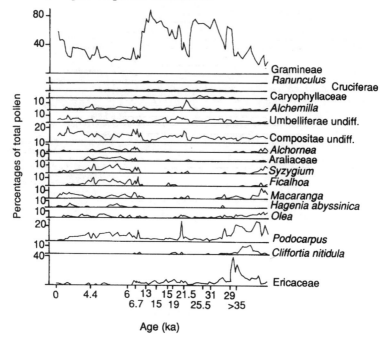

Figure 5.22 Abundances of pollen types in the late-Quaternary sediment sequence from Kashiru, Burundi. Redrawn from Bonnefille & Riollet (1988, Fig. 6).

subsequently dried out and their sediments have been buried by aeolian sand (Fig. 5.23). The sites of Oyo (Ritchie *et al.* 1985), Bir Atrun (Ritchie 1987b; Ritchie & Haynes 1987), and Selima (Ritchie & Haynes 1987; Haynes *et al.* 1989) are all now in the hyper-arid zone (*sensu* White 1983) of the Sahara in a landscape of vast extents of sand and rock and only occasional plants (grasses: *Stipagrostis*; shrubs: *Capparis decidua, Salvadora, Maerua crassifolia*) of Saharo–Mediterranean floristic affinity. Individuals of *Capparis decidua* may be separated by about 100 km. Towards the south, annual rainfall increases, and desert scrub and grassland develops, becoming thorn savanna and then deciduous savanna woodlands where the annual rainfall exceeds about 600 mm (Ritchie 1987b). Pollen spectra at all sites are dominated by a few wind-pollinated types (Gramineae and Cyperaceae), so Ritchie *et al.* (1985) based their interpretations on the recognition of small numbers of pollen grains from tropical taxa with low pollen productivity (*Piliostigma* and *Grewia tenax*) and low dispersal capacity because of large size (*Acacia* and *Grewia tenax*).

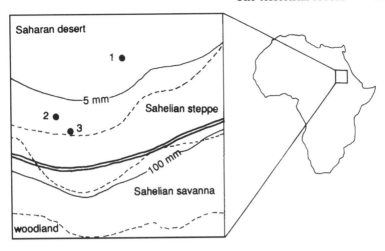

Figure 5.23 Location of pollen sites in the arid core of the eastern Sahara. 1, Selima (see Fig. 5.24); 2, Oyo; 3, Bir Atrun. Labelled contours show annual rainfall, and the thick double line indicates the northern extent of the monsoon. Named vegetation zones are separated by the dashed lines. Redrawn from Ritchie & Haynes (1987, Fig. 1).

Sediments at the most northerly site, Selima, dated to between 8.4 ka and <6 ka, include pollen spectra dominated by desert–grassland types. At about 8 ka, sediments indicate that a deep (20 m) lake existed, and the pollen record, with maximum values for Sahelo–Sudanian types, indicates a thorn tree–savanna vegetational mosaic with *Commiphora*, *Acacia* and *Capparis* (Fig. 5.24). Haynes *et al.* (1989) suggested, cautiously, that vegetation zones may have been displaced 4–5° latitude northwards, with more than 300 mm annual precipitation at Selima. Further south at Oyo, sediments dated between 8.5 ka and 6.1 ka suggest the existence then of a deep, stratified lake. Pollen records indicate that it was surrounded by continuous savanna woodland with taxa such as *Grewia tenax* and *Piliostigma*, now found about 500 km further south. This would have needed a tropical monsoonal climate with annual rainfall of at least 400 mm. Sediment and pollen analyses indicate declining annual rainfall to less than 100 mm by 4.5 ka. Bir Atrun is slightly further south, near the present southern limit of absolute desert, with lake sediments dated to between 9.6 ka and 6.2 ka. Pollen data show similar high frequencies for Sudanian savanna types, and lower frequencies for the Saharo–Mediterranean element than Selima. As at Oyo, these records

130 · Biological response: distribution

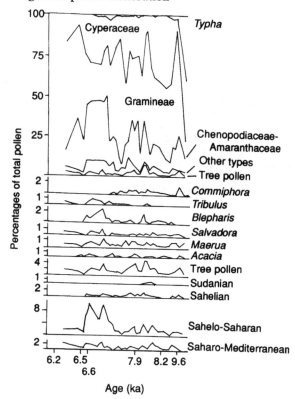

Figure 5.24 Abundances of pollen types in Holocene lake sediments from Selima, Sudan (see Fig. 5.23). From the base upwards, figures shown are pollen types grouped phytogeographically, followed by the sum of all tree pollen, the abundances of the important tree pollen types, and finally a summary of all pollen taxa. Redrawn from Ritchie & Haynes (1987, Fig. 2a).

indicate the existence of continuous savanna woodland during the early Holocene (Ritchie 1987b; Ritchie & Haynes 1987).

The existence of these sites is, first of all, evidence for more northerly extension of the east African monsoon during the Holocene than today, confirming indications of computer modelling based on orbital forcing (Kutzbach & Otto-Bliesner 1982; COHMAP members 1988; Kutzbach *et al.* 1993). The humid period that resulted displaced vegetation belts northwards for the duration of the phase (less than 5 ka). There is as yet insufficient palaeoecological data from this region to assess the independence of change in the distributions of the species concerned, and

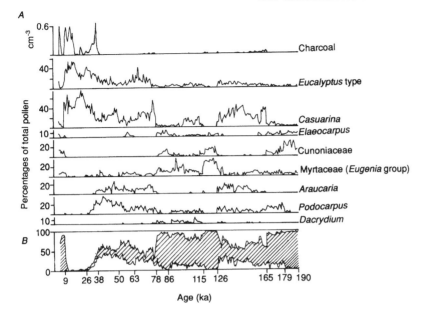

Figure 5.25 A. Abundances of pollen types and charcoal in Quaternary lake sediments from Lynch's Crater, northeastern Australia.
B. Combined abundances of pollen types in the following groups: rain-forest gymnosperms (*Dacrydium*, *Podocarpus*, and *Araucaria*) in the lowermost curve; then rain-forest angiosperms (*Eugenia*-group, Cunoniaceae, and *Elaeocarpus*), shaded; then sclerophylls (*Casuarina*, and *Eucalyptus*-type) in the uppermost curve. Redrawn from Kershaw (1986, Fig. 1).

insufficient data from modern pollen spectra to assess whether displaced vegetation included communities different from modern or not. Nevertheless, it is clear that trees and shrubs shifted their distributions by several hundred kilometres in response to a change in monsoonal extent, and this provides an illuminating example of the way plant populations can respond to orbitally-forced climatic change in tropical latitudes where the dominant limiting climatic parameter is not temperature.

Australasia
Two long pollen records provide information on changing vegetation in eastern Australia during multiple climatic oscillations. The modern vegetation of this part of the continent consists mostly of *Eucalyptus* woodland, with patches of rain forest, ranging from cool temperate in the south to tropical in the north, where rainfall is seasonally high.

Lynch's Crater, at an altitude of about 650 m on the Atherton Tableland of northeast Queensland, is a volcanic crater that has infilled with about 60 m of lake and swamp sediments. A pollen record from the entire sequence is described by Kershaw (1985, 1986). It covers the last two climatic oscillations (Fig. 5.25), using a time-scale provided by radiocarbon age determinations for the most recent 40 kyr, and by comparison with other evidence, especially $\delta^{18}O$ stratigraphy from deep-sea sediments, for the rest of the sequence (Kershaw 1978, 1985). The main changes involve shifts in the relative proportions of three groups: sclerophylls (for example *Casuarina* and *Eucalyptus*), rain-forest angiosperms (for example Cunoniaceae, Myrtaceae (*Eugenia*-group), and *Elaeocarpus*), and rain-forest gymnosperms (for example *Dacrydium*, *Podocarpus*, and *Araucaria*). Broadly, the record shows the development of angiosperm rain forest during inferred high-precipitation 'interglacial' periods (190–179 ka, 126–115 ka, and since 9 ka), with development of rain-forest gymnosperms and sclerophylls during low-precipitation 'glacial' periods. However, this generalization overlies more complex behaviour from some species. Thus, the representation of Myrtaceae (*Eugenia*-group) is inconsistent between the three 'interglacial' periods, and the rain-forest gymnosperms reach higher frequencies in the upper 'glacial' phase (78–38 ka). Between 115 ka and 78 ka, there are high frequencies of *Dacrydium*, and also pollen of the gymnosperm *Phyllocladus* and the rain-forest angiosperm group *Nothofagus brassii*-type. All three taxa are now extinct on mainland Australia, but do occur on New Guinea and (except the *Nothofagus*) Tasmania and New Zealand. This period in the record, at least, must represent a vegetation type for which there is no modern analogue in continental Australia. The pollen spectra at 190–179 ka appear also to represent vegetation for which there is no modern analogue.

The shift towards dominance by sclerophylls towards the end of the most recent 'glacial' (26–9 ka) is parallelled by high values for microscopic charcoal in the sediments (Fig. 5.25), suggesting strongly that the increase of sclerophylls then is due to an increase in woodland burning. Kershaw (1985, 1986) argued that this may have resulted from burning by aboriginal people (see also Singh, Kershaw & Clark 1981).

Lynch's Crater is located within the tropics, in an area of high, but seasonal, rainfall, and remote from any possible direct influence of Quaternary glaciers. Kershaw (1985, 1986) interpreted the sequence mainly in terms of changing rainfall, which is the most likely controlling influence of the forest changes in the area resulting from the Earth's orbital variations. This site is an important record of vegetation change involving entire forest types (see, for example, the transitions at 126 ka and 86 ka) on time-scales identical to those found in high-latitude records, in an area of probable low-amplitude, orbitally-forced climatic change (Kutzbach & Guetter 1986; Kutzbach et al. 1993).

The second long record from eastern Australia comes from the large tectonic basin of Lake George, in New South Wales (Singh & Geissler 1985). It is an extant lake with a fluctuating water level probably controlled by the balance between rainfall and evaporation. The lake is within the region in which *Eucalyptus* woodland predominated before European settlement. A 72-m core of sediment has been recovered from the lake bed, of which the upper 18 m is polliniferous (Fig. 5.26). The sequence has been dated by radiocarbon age determinations at the top, and palaeomagnetic stratigraphy, including the Brunhes–Matuyama magnetic reversal (730 ka), for the rest of the sequence (Singh, Opdyke & Bowler 1981). About eight 'glacial–interglacial' oscillations occur during the 730 kyr of the Brunhes Chron and have been correlated with stages 1–19 of the deep-sea $\delta^{18}O$ record of Shackleton & Opdyke (1973). 'Interglacials' are dominated by woodlands of the sclerophyll *Casuarina*, together with 'wet-sclerophyll' forest taxa, until the two most recent oscillations, when *Eucalyptus* became more prominent, reaching its modern dominance in the complete absence of *Casuarina*. 'Glacial' periods are varied and marked by herbs, suggesting open, non-forest vegetation. At intervals, pollen of cool-temperate trees such as Podocarpaceae and *Nothofagus* is found, sufficient to suggest that some temperate woodland may have grown near the lake, especially during transitions from sclerophyll woodland to open vegetation. A number of the cool-temperate trees, such as the podocarp trees *Microcachrys*, *Phyllocladus*, *Dacrydium*, *Podocarpus*, and three groups of *Nothofagus* (*menziesii*-type, *fusca*-type, and *brassii*-type) all disappear from the record during the Brunhes Chron. At least some of these extinctions occurred late enough to be related to a general rise in burning evident from the microscopic charcoal record (Fig. 5.26). They coincided with the increase in *Eucalyptus* at the expense of *Casuarina* as the dominant of 'interglacial' sclerophyll woodland, which is possibly attributable to anthropogenic woodland burning as early as 130 ka (Singh & Geissler 1985).

The record at Lake George, like Lynch's Crater, reveals patterns of vegetation change in middle and low latitudes at least as dramatic as changes nearer to centres of continental glaciation. There is alternation, at a broad level, of sclerophyll forest with various open plant communities, superimposed on longer-term shifts due to extinctions, particularly of

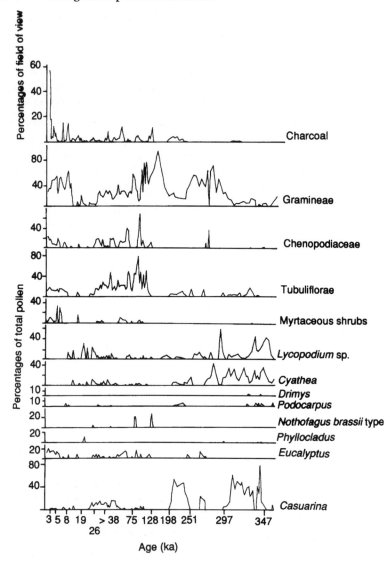

Figure 5.26 Abundances of pollen and spore types and charcoal in Quaternary lake sediments from Lake George, eastern Australia. Charcoal abundances are expressed as percentages of microscopic fields of view covered by charcoal, for volumetrically standardized samples. Redrawn from Singh & Geissler (1985, Figs 11 and 12B).

cool-temperate trees, and changes that may be associated with burning by aboriginal people.

Late-Quaternary vegetation dynamics of New Zealand are reviewed by McGlone (1988). Vegetation at the last glacial maximum was predominantly herbaceous, except in the far north of North Island (Dodson et al. 1988), in contrast to the probable almost completely forested state at the time of Polynesian occupation about 1000 years ago (McGlone 1989; Anderson & McGlone 1992). The timing of reafforestation varied across the islands within the period 14–9.5 ka, but was always an abrupt change, taking place within a few hundred years. Environmental gradients appear to have controlled the type of forest that appeared within any region, but did not influence rates of spread or timing. McGlone (1985, 1988) argued that populations of forest tree species survived the glacial maximum in many, widely separated areas, and the populations may have been small. These then provided the nuclei for increase and establishment of new, larger populations, leading to regional reafforestation. There seems to be little evidence for time-transgressive spread of species distributions in New Zealand (but see Wardle (1988) for a different view).

Pollen records from low altitudes in the tropics are scarce, but van der Kaars (1991) has obtained a series of late-Quaternary pollen sequences from marine piston cores in the eastern Indonesia–New Guinea–Australia area. Their interpretation is not straightforward because of problems with pollen source area and changing sea-levels, but his results do indicate that grasslands were more extensive in northern Australia during glacial periods, and forest and fern vegetation increased during interglacials, including the Holocene. Phases with high frequencies of mangrove pollen mark changing sea-levels at the sites he investigated. This gives some insight into a complicated set of vegetation transformations in this region. Orbitally-forced climatic changes caused transformations of forest to grassland, and back, simultaneously with major sea-level changes, caused by the waxing and waning of ice-sheets, which radically altered the amount of land available for colonization by terrestrial plants, and enabled a back and forth movement of coastal mangrove across the region.

A mire at Lake Hordorli, at about 780 m altitude in the north coastal range of New Guinea, provides evidence for vegetational dynamics in the area over the last 60 ka (Hope & Tulip 1994). The pollen record (Fig. 5.27) is readily divided into a section dominated by pollen of *Nothofagus* before about 7 ka, and an upper section with lower values for *Nothofagus* and high frequencies for pollen from plants associated with forest

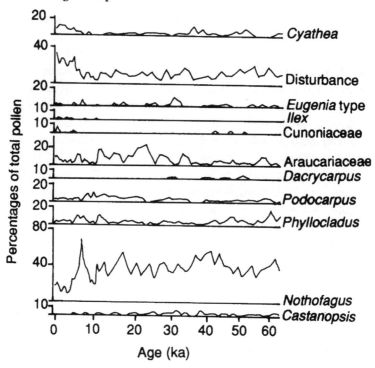

Figure 5.27 Abundances of pollen and spore types in late-Quaternary lake sediments from Lake Hordorli, New Guinea. Redrawn from Hope & Tulip (1994, Fig. 4).

disturbance, reflecting anthropogenic activity near the site. Frequencies of *Nothofagus* pollen fluctuate in the period before 7 ka, with the more abundant taxa being those now characteristic of higher altitudes (for example Araucariaceae) between about 10.5 ka and 25 ka, and higher frequencies of the podocarp trees *Dacrycarpus* and *Phyllocladus* earlier on. The vegetation around the site has clearly been forest-dominated for at least the last 60 kyr, but with fluctuations in its composition that appear to be related to altitudinal shifts of the various components of the region's montane forests (Hope & Tulip 1994). Fluctuations in forest composition at about 10.5 ka are especially notable as evidence of the potential rapidity of forest change.

Asia

The Quaternary pollen record from Asia is, unfortunately, sparse. Projects underway on late-Quaternary vegetation and environmental dynamics in Tibet (van Campo & Gasse 1993; van Campo *et al.* 1996) and Siberia (Lozhkin *et al.* 1993) hold considerable potential as the basis for continental-scale reconstructions, but the density of suitable sites is not yet adequate. An exceptionally long sequence of sediments from the base of Lake Biwa in Japan (Fuji 1983, 1986) holds considerable promise for a detailed record of changes in Japanese vegetation over the last 600 kyr when analyses are completed and available. Tsukada (1988) has reviewed Japanese late-Quaternary vegetation dynamics. At the last glacial maximum, the archipelago was largely forested, but dominated by coniferous elements. Forests with a warm temperate character were probably developed only on the palaeo-Yaku peninsula, in the extreme south of the archipelago and on a part of the continental shelf now submerged (Tsukada 1985). Modern temperate broadleaf forests developed during the Holocene. Coniferous forests disappeared from the lowlands, and *Larix gmelinii* became extinct in the archipelago at about 10 ka. After about 4 ka, subalpine conifers extended down the mountains to their present limits, while *Cryptomeria japonica* populations became established further north, reaching its modern limit at 4 ka. There has thus been a complete change in the composition of Japanese forests during the last 20 kyr.

In Indonesia, sediment cores from the intra-montane Bandung basin, western Java provide a record spanning the last 135 kyr, giving the first indication of vegetation and environmental change over the last glacial–interglacial oscillation from the region (van der Kaars & Dam 1995). During the transition into full-glacial conditions, broadly-defined forest types shifted downwards in altitude by about 600 m. Pollen of *Ulmus*, now absent from Java, was present within the last glacial period, indicating a change in the composition of these forest types.

Animals

Beetles
Fossil beetles are often found in abundance in many unconsolidated organic deposits, and specimens can be identified to species using the same morphological features that would be used to identify modern specimens. Coope (1986) and Elias (1994) outline methods and procedures. However, analyses of suitable deposits have so far been largely limited to western Europe and North America.

Coope (for example 1970, 1979, 1995) has drawn attention to some of the vast distributional changes among British beetle faunas that have taken place in response to Quaternary climatic changes. Beetles appear to have responded quickly to climatic change, so beetle populations became established up to 1 kyr more rapidly than populations of trees following a climatic warming (Coope 1977). The most abundant dung beetle in fossil assemblages of the middle of the last cold stage is *Aphodius holdereri* (Coope 1973). Today, this species is restricted to the high plateau of eastern Tibet at altitudes of 3000–5000 m. Beetles associated with it during the last cold stage in Britain include a number of species with present distributions in eastern Asia. One of these, *Helophorus aspericollis*, now found in eastern Siberia, was abundant in Britain during the last cold stage, but is absent today (Angus 1973). Its congener, *Helophorus brevipalpis*, occurred in Britain during warm stages. The distributions of the two today suggest a formerly continuous distribution that had been separated by the expansion of the Fennoscandian ice-sheet, but the fossil record shows that both forms are extremely mobile, and their present geographical locations have more to do with establishment in environments that each finds to be suitable, and nothing at all to do with evolutionary history (Coope 1979).

At different times during the last glacial–interglacial oscillation (since about 125 ka), the British beetle fauna has included assemblages typical of modern southern Britain, southern Europe, boreal forests, and high arctic faunas, often with modern Siberian species (Coope & Angus 1975: and see Fig. 5.28). But although each of these faunas has general resemblances to the areas indicated, there are nearly always exceptions. The fauna of a mid-last-glacial warm stage near London (Coope & Angus 1975) is predominantly typical of southern England today. However, it includes species with modern distributions in southern Spain and north Africa (for example the curculionid *Cathormiocerus curviscapus*) and species found today in eastern Europe, including Scandinavia (for example the carabid *Calosoma reticulatum*).

A sequence of late last-glacial lacustrine silts in southern Minnesota has yielded a rich insect assemblage (almost all beetles) dating to between 12.4 ka and 11.2 ka (Ashworth *et al.* 1981). The beetle fauna includes species of modern boreal forest, tundra–forest, and western distributions, and has no modern analogue: there is nowhere today where all the species found co-exist (Fig. 5.29). Morgan & Morgan (1980) illustrated many transcontinental North American distribution changes involving, especially, species of beetle found as fossils during the last glacial in the Great Lakes region but which are known today only from northwestern

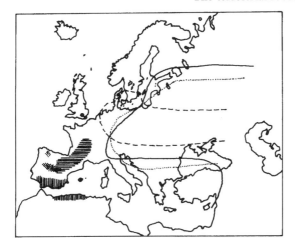

Figure 5.28 Intermingled fauna: European beetles. Present distributions are shown as follows: vertical lines for *Aphodius bonvouloiri* (Scarabaeidae); horizontal lines for *Cathormiocerus curviscapus* (Curculionidae); stipple for *Cathormiocerus validiscapus* (Curculionidae); continuous line for western limit of *Onthophagus gibbulus* (Scarabaeidae); dashed line for western limit of *Calosoma reticulatum* (Carabidae); dotted line for western limit of *Opetiopalpus scutellaris* (Cleridae). All occur at the full-glacial site at Isleworth, England, marked by a filled circle. Redrawn from Coope & Angus (1975, Figs 4 and 5).

North America (Alaska and adjacent Canada). For some species (for example the carabid *Asaphidion yukonense* and the scolytid *Carphoborus andersoni*), there are modern localities well within the area covered by the Laurentide ice-sheet during the last glacial, so there have certainly been distribution changes due to the movement of the insects and not just a formerly extensive distribution broken up by the last glacial.

Beetle communities did not react *en bloc* to Quaternary climatic changes. Each species responded individualistically, and at its own rate, leading to the development of novel species associations lasting for thousands of years while various species' distributions adjusted to new climatic conditions following, for example, the rapid ending of a cold stage (Coope 1987).

Non-marine molluscs
The record of non-marine molluscs has been particularly well-studied in terrestrial and freshwater deposits of the British Isles. The Holocene fauna is largely an immigrant fauna, arriving sequentially because of the

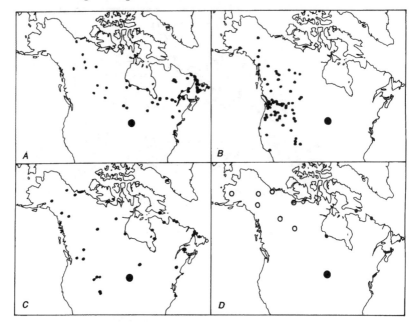

Figure 5.29 Intermingled fauna: North American beetles. Figures show modern distributions of:
A. *Pterostichus punctatissimus* (Carabidae).
B. *Opisthius richardsoni* (Carabidae).
C. *Cymindis unicolor* (Carabidae).
D. *Helophorus arcticus* (Hydrophilidae: filled circles) and *Carphoborus andersoni* (Scolytidae: open circles).
All occur today at the last glacial site at Norwood, Minnesota, marked by a large filled circle.
Redrawn from Ashworth *et al.* (1981, Fig. 4).

time needed for populations to spread northwards following the end of the last glacial, either because of the dependence of the molluscs on vegetation development, or because of their climatic requirements in relation to the pattern of climatic change at the transition of the beginning of the Holocene (Kerney *et al.* 1980). Faunas of the late-glacial and early Holocene have no modern counterparts (Fig. 5.30).

Gould (1970) described Quaternary fluctuations in abundance for the Bermudan land-snail species *Gastrocopta rupicola* and *Thysanophora hypolepta*, more frequent during interglacials and glacials respectively. The Bermudan land-mass is sufficiently small that we can be reasonably sure that these are real changes in species relative abundance, and not a consequence of movements (see also page 161).

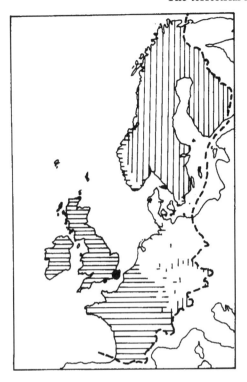

Figure 5.30 Intermingled fauna: North European land mollusca. Vertical shading and horizontal shading gives the modern distributions of *Discus ruderatus* (Endodontidae) and *Lauria cylindracea* (Pupillidae), respectively, in the region north and west of the dashed line. Both species occurred during the early Holocene in southern England (for example at Folkestone, marked by a filled circle). From maps in Kerney & Cameron (1979, pp. 234 and 237) and data in Kerney *et al.* (1980).

Vertebrates

The late-Quaternary fossil record for North American mammals has been reviewed by Lundelius *et al.* (1983), Semken (1984), and Graham (1986), and detailed maps are available in the FAUNMAP database (Graham & Lundelius 1995). Nearly all faunas of the last glacial are 'disharmonious' (Lundelius *et al.* 1983), consisting of species that are today widely separated geographically, and which seem to be ecologically incompatible. Faunas approached the modern situation during the Holocene, with distributional changes that involved movements of species from all ecosystems moving in all possible directions (Semken 1984). Figs 5.31

142 · Biological response: distribution

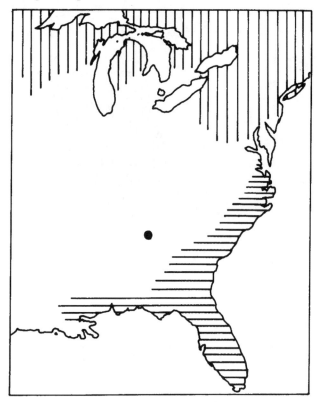

Figure 5.31 Intermingled fauna: North American vertebrates. Vertical shading gives the modern distribution of *Clemmys insculpta* (wood turtle), and horizontal shading gives the modern distribution of *Bufo terrestris* (southern toad). Both occur at the late-Quaternary site at Ladds, Georgia, marked by a filled circle. Redrawn from Holman (1976, Fig. 1).

and 5.32 illustrate North American herpetological and mammal community changes since the last glacial for combinations of species whose distributions overlapped then, but not today. Similar intermingled faunas are also found elsewhere in the world during the last glacial (Fig. 5.33). Lundelius (1983) has detailed species pairs among Australian last glacial assemblages that are now allopatric. Faunas more recent than about 16 ka have fewer disharmonious species and suggest changes in distribution towards those of the present day. Lundelius *et al.* (1983), Semken (1984), Graham (1986), and Graham & Grimm (1990) have all commented on the individual nature of vertebrate community reorganization since the last glacial.

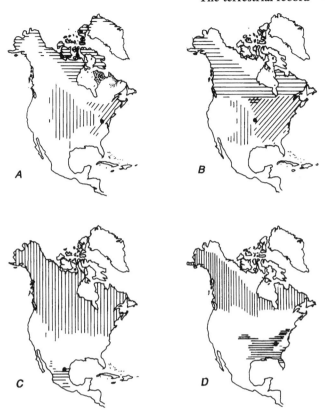

Figure 5.32 Intermingled fauna: North American mammals. In each figure, the species whose modern distributions are shaded occurred at the late-Quaternary sites indicated by filled circles.

A. *Sorex fumeus* (smoky shrew: diagonal shading), *Spermophilus tridecemlineatus* (thirteen-lined ground squirrel: vertical shading), *Dicrostonyx hudsonius* (Hudson Bay collared lemming: stipple), and *Dicrostonyx torquatus* (collared lemming: horizontal shading). The first two and a species of *Dicrostonyx* occurred in Pennsylvania during the late-glacial.

B. *Synaptomys borealis* (northern bog lemming: horizontal shading), *Tamias striatus* (eastern chipmunk: diagonal shading), and *Cynomys ludovicianus* (blacktail prairie dog: vertical shading) occurred in the full-glacial of Iowa.

C. *Sorex cinereus* (masked shrew: vertical shading) and *Liomys irroratus* (Mexican pocket mouse: horizontal shading) occurred in the glacial of Mexico.

D. *Neotoma floridana* (eastern woodrat: horizontal shading) and *Synaptomys borealis* (vertical shading) occurred in the late-glacial of Tennessee.

Redrawn from Graham (1986, Fig. 18.3).

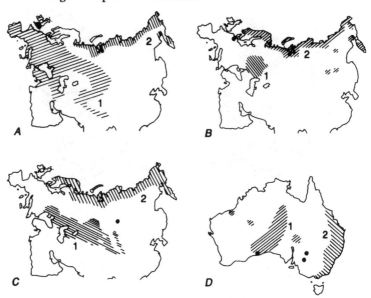

Figure 5.33 Intermingled fauna: Eurasian and Australian mammals. In each figure, the species whose modern distributions are shaded both occur at the late-Quaternary sites indicated by filled circles.
A. *Apodemus sylvaticus* (wood mouse) (1) and *Dicrostonyx torquatus* (collared lemming) (2) both occur at full-glacial sites in England.
B. *Desmana moschata* (Russian desman) (1) and *Lemmus lemmus* (Norway lemming) (2) both occur at full-glacial sites in northern Germany.
C. *Saiga tatarica* (saiga) (1) and *Alopex lagopus* (arctic fox) (2) both occur at full-glacial sites in Siberia.
D. *Dasycercus cristicauda* (mulgara) (1) and *Phascolarctos cinereus* (koala) (2) both occur at full-glacial sites in southern Australia.
Redrawn from Graham (1986, Fig. 18.4).

Riddle et al. (1993) used details of mitochondrial DNA phylogenetic structure in grasshopper mice (*Onchyomys leucogaster*) to evaluate the responses of these mammals to Quaternary climatic oscillations in arid regions of western North America. Their data indicate that grasshopper mice must have persisted within intermontane regions during at least the last glacial oscillation. It is not clear whether their distributions shifted along altitudinal gradients (as plants did), but there is certainly the possibility that certain species survived major climatic oscillations *in situ*.

The record of deer in the British and European Quaternary shows considerable mobility in a group where species seem to have had short durations (Lister 1984a, 1986). Reindeer (*Rangifer tarandus*), now restricted to arctic regions, ranged south during late-Quaternary cold stages right across Britain and continental Europe as far south as northern Spain (Lister 1986). No British interglacial fauna contains any reindeer remains, suggesting that distribution changes of this animal were extensive and comprehensive between glacial and interglacial stages. Conversely, other deer (for example fallow deer (*Dama dama*) and roe deer (*Capreolus capreolus*)) are known only from interglacials in Britain. Fallow deer spread into Britain, and were common, during two previous interglacials, presumably from glacial distributions in southern Europe, but have not spread 'naturally' beyond the Mediterranean region during the present (Holocene) interglacial. Red deer (*Cervus elaphus*) may have persisted in northern Europe throughout the late-Quaternary, but with much reduced population sizes during glacials (Lister 1984a, 1986).

The marine record

The nature of the Quaternary palaeontological record from the oceans differs from the terrestrial record and the pre-Quaternary marine record. Fossiliferous sediment is collected almost entirely by coring operations, so usually the only fossils available are microfossils. The best studied are unicellular algae and protozoans with siliceous or calcareous tests, including both planktonic and benthic forms. These organisms have short generation times and have responded rapidly (instantaneously for all practical purposes) relative to the climatic changes of the Quaternary, unlike the changing distributions of long-lived trees, which may lag behind significant climatic changes. Since extensive sections and exposures of Quaternary marine sediment are rare, relatively little is known of the Quaternary fossil record of large-bodied marine organisms.

Microfossils

Most of our knowledge of Quaternary climatic change has come from work on marine microfossils. As ocean waters warmed and cooled during the Quaternary, planktonic and benthic organisms shifted location and changed in abundance on a massive scale. Although planktonic organisms could move hundreds of kilometres in one year by advection and diffusion in water masses, the evidence is that they do not (Dexter *et al.* 1987). Areas with large numbers of a particular species remain fixed in location for hundreds of years, moving only slowly over thousands of years as

146 · Biological response: distribution

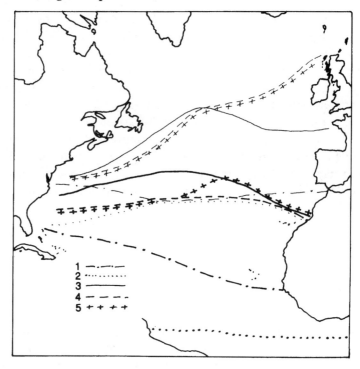

Figure 5.34 Changing northern limits for species of coccolithophores in the North Atlantic between full-glacial and modern times. 1, *Coccolithus pelagicus*; 2, *Umbellosphaera irregularis*; 3, *Helicospaera carteri*; 4, *Rhabdosphaera stylifera*; 5, *Syracosphaera pulcha*. The thick line applies to full-glacial distribution, the thin line to modern distribution. Redrawn from McIntyre (1967, Fig. 1).

water masses change in temperature, salinity, or some other aspect of their composition. Benthic organisms also 'move' as temperature and other environmental conditions of the ocean floor are affected by Quaternary climatic fluctuations.

The shifting distributions of several species of coccolithophores in the North Atlantic are described by McIntyre (1967). Each reacted slightly differently to the same external change in ocean conditions, some by means of distribution changes north and south (Fig. 5.34), and others (for example *Cyclococcolithus leptoporus*) through abundance changes within an overall distribution that changed little from glacial to modern conditions. In the subantarctic Pacific Ocean, over the last 400 kyr, coccolithophore floras have changed from domination by *Gephyrocapea caribbeanica* in glacial-age sediments to domination by *Cyclococcolithus lepto-*

porus and *Coccolithus pelagicus* in interglacial sediments (Geitzenauer 1972). However, *Gephyrocapea caribbeanica* became dominant during a period of intense warming (based on foraminiferal temperature reconstructions) at about 400–500 ka. In the Atlantic today, north of latitude 40°S, the benthic foraminifer *Uvigerina peregrina* is found only at depths of less than 2 km (Streeter & Shackleton 1979). During the last glacial it was abundant at depths of 3.3 km at 44°N. The distribution change between then and now appears to be due to an increase in the North Atlantic Deep Water, bringing cold, oxygenated water at considerable depth from the arctic. The radiolarian *Cycladophora davisiana*, has a cosmopolitan distribution in the ocean, but is now abundant only in the Sea of Okhotsk, apparently due to the development there of intense winter sea-ice that melts completely in summer, giving rise to a thick layer of cold, low-salinity water (Morley & Hays 1983). *Cycladophora davisiana* has, however, been abundant in several parts of the ocean of both hemispheres during glacial phases of the Quaternary, showing a globally-synchronous population change from the North Atlantic Ocean and northwest Pacific Ocean to the subantarctic in response to Quaternary climatic changes: abundant during cold stages, rare (except locally) in warm stages (Morley & Hays 1979; Morley *et al.* 1982). McIntyre *et al.* (1972) described southward movements of coccolithophores and foraminifera in the North Atlantic over the past 225 kyr. During glacial phases, northern populations of some species of foraminifera moved far enough south to become contiguous with southern populations spreading north. Imbrie & Kipp (1971) discussed the movement of subpolar foraminifera into the Caribbean during the last glacial. Marine diatom floras also responded to Quaternary climatic shifts by large changes in relative abundances. In the Bering Sea during the last glacial and the Holocene, the main species have been present throughout, but have fluctuated in relative abundances (Sancetta & Robinson 1983). Thus, *Denticulopsis seminae* was dominant during the Holocene, but *Thalassiosira graviola* was dominant during the last glacial.

Molluscs

Among marine macrofauna, the molluscan record is particularly well-known. Raffi (1986) documented several incursions of northern molluscs into the Mediterranean during the early Quaternary. The bivalve *Arctica islandica*, which today does not occur south of the English Channel, co-existed in the Mediterranean during the early Quaternary with the modern Mediterranean bivalves (*Chlamys flexuosa*, *Chlamys glabra*, *Pecten jacobaeus*, *Spondylus gaederopus*, *Tellina compressa*, and *Tellina pulchella*) that today do not occur north of the coast of Portugal. Thomsen & Vorren (1986) and Peacock (1989) described the movements of molluscs along northeastern North Atlantic coasts during the late-Quaternary as glacial-age arctic faunas of Britain and northern Norway were replaced by members of the modern fauna spreading from the south. Quaternary sea-level fluctuations had virtually no effect on Californian marine molluscan fauna (Valentine & Jablonski 1991). Periods of high sea-level show continuity of communities (for example coastal marsh, and rocky shore) and species composition over the last million years. The most significant biotic changes were associated with latitudinal shifts in species' geographic ranges, forced by climatic changes.

Corals

Potts (1983, 1984) discussed the persistence of reef-building corals in the Indo-Pacific regions. He has shown (Fig. 5.35) that late-Quaternary sea-level changes, occurring at rates of 25 mm yr^{-1}, shifted the growth zone of shallow-water corals laterally at rates of up to 50–100 m yr^{-1} on the Sunda and Sahul shelves of southeast Asia and northern Australasia. Areas of the shelves within the overall altitudinal range of sea-level fluctuations were within the growth-zone for these corals for, on average, just 3.2 kyr at a time. Reef-building corals must have matched the rate of marine transgression to have survived, presumably by colonizing new habitats.

Discussion

West (1964), through comparison of floras of different interglacials, appreciated that communities are only 'temporary aggregations', and he thus provided an early insight into the discoveries of the following three decades. This discussion was, however, foreshadowed by the interpretation of Gleason (1926) concerning the nature of plant communities. The results reviewed here are a major scientific development of those insights, developed crucially from the application of the radiocarbon method of age determination after the 1970s. The analysis of Quaternary fossil sequences has been expanded and improved considerably, but the method of obtaining ages for each sequence studied has enabled an appreciation of the rates of change of distribution and abundances. In particular, it

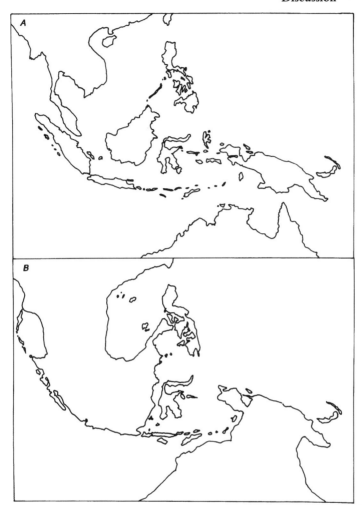

Figure 5.35 Changing Quaternary coastlines between south-east Asia and northern Australia.
A. Modern configuration.
B. Configuration with sea-level 130 m lower than today, at the probable minimum level of the last glacial maximum.
Redrawn from Potts (1983, Fig. 3).

is now clear that break-up of communities because of the individualistic response of species to orbitally-forced climatic oscillations is a global phenomenon, involving tropical communities as well as temperate and high-latitude systems (Flenley 1979), marine as well as terrestrial communities. The view of many ecologists (for example Wilson 1994) that the tropics have been a centre of community stability for millions of years is no longer tenable.

Movement is clearly an important biological response to Quaternary climatic changes, but it is necessary to be clear about exactly what it is that is moving in the examples discussed above. In the case of forest trees on the northern continents, new northern limits for members of a species during an interglacial are established many generations removed from glacial ancestors. The only pieces of living tissue that move any distance are seeds (with or without encasing fruit material) which move only kilometres, at most, and pollen, which may be dispersed further. Similarly, no single beetle was transported from Britain to Tibet during the last cold-stage. Populations do not move, and communities disintegrate. The only category that can be said to have moved is the species. Distributions are thus emergent properties of species, forming the spatial aspect of the claim for species to be separate entities, or 'individuals' (see page 30). Distribution changes in response to glacial–interglacial oscillations are the consequence of climatic change forcing species into new spatial configurations.

Distribution change of species on subcontinental and continental scales involves, potentially, an enormous increase in the numbers of individuals comprising the species concerned. In Europe and eastern North America, for example, several tree species increased from low abundances in scattered sites within a small area of the south of either continent to become forest dominants across most of the region. The total numbers of individuals for these species might have increased by several orders of magnitude within the first few thousand years of the Holocene, and such changes may have been typical behaviour over many earlier interglacials. Following a suggestion by Watts (1973), investigation at a number of individual sites shows that build-up of populations in any one area is, initially at least, geometric for Holocene forest taxa, in temperate and tropical regions (Table 5.1) and for interglacial temperate taxa (Table 5.2). Doubling times for population increase are of the order of a few tens to a few hundred years, and are similar to those obtained from modern tree populations over much shorter periods of time (Bennett 1986b; Prentice 1988; MacDonald 1993). These rates of increase are maintained

Table 5.1. *Doubling times estimated for the exponential increase of pollen taxa during the early Holocene*

Taxon	Doubling time	Region	Sources
Abies	408	N. America	3
Acer saccharum	300, 350	N. America	9
Agathis	199	Australia	7
Alnus glutinosa	174, 268	England	4, 5
Alnus rubra	112, 178	N. America	3
Balanops	171	Australia	7
Betula	59	England	4
Corylus avellana	45, 46	England	4, 5
Cunoniaceae	118	Australia	7
Elaeocarpus	106	Australia	7
Eugenia	355	Australia	7
Fagus grandifolia	256, 444, 470, 780	N. America	6, 9
Fagus	187	Japan	2
Fraxinus nigra	200, 290, 380	N. America	9
Carpinus/Ostrya	280, 370	N. America	9
Picea glauca	770	N. America	9
Picea	35, 178	N. America	3
Pinus contorta	107, 173, 80–1100	N. America	3, 8
Pinus strobus	150, 240, 580	N. America	9
Pinus sylvestris	73	England	4
Podocarpus	236	Australia	7
Pseudotsuga menziesii	52, 171, 365	N. America	1, 3
Quercus	78, 141	England	4, 5
Rapanea	150	Australia	7
Salix	204	N. America	3
Thuja plicata	224, 231, 239	N. America	3
Tilia cordata	99	England	4
Trema	78, 107	Australia	7
Tsuga canadensis	200, 320, 450	N. America	9
Tsuga mertensiana	31, 462	N. America	3
Ulmus	29, 67	England	4, 5
Ulmus	60, 170, 500	N. America	9

Note: Doubling times are in radiocarbon years.
Sources: 1, Tsukada (1982a); 2, Tsukada (1982b); 3, Tsukada & Sugita (1982); 4, Bennett (1983); 5, Bennett (1986a); 6, Bennett (1988); 7, Chen (1988); 8, MacDonald & Cwynar (1991); 9, Fuller (1995).

for periods of time measured in hundreds or even thousands of years, and thus involve massive increases of population size of many orders of magnitude. The transition into interglacial conditions seems to provide opportunities for species to increase in abundance geometrically, occupying and dominating considerable areas in the process. This unchecked

Table 5.2. *Doubling times estimated for the exponential increase of pollen taxa during late-Quaternary interglacials at Valle di Castiglione, Italy*

Taxon	Doubling time (years)
Abies	1117
Carpinus	369, 438, 1362
Corylus	438
Fagus	368, 811
Juniperus	1177
Olea	621
Pinus	1128
Quercus (deciduous)	1041, 1289
Tilia	720
Ulmus	664
Zelkova	713

Source: From Magri (1989, Table 1).

population growth is just the situation Elton (1927) envisaged as being necessary to enable the spread of non-adaptive characters through populations (see pages 36–37).

The action of climate on species distributions is, possibly uniquely, a genuine example of species acting as interactors (*sensu* Hull 1980), and the single most important factor in enabling speciation. Distributions are pushed and pulled, this way and that, again and again. Eventually, something gives. A bit of the distribution becomes separated, and the two pieces may then have different histories, including possible development as new species. How long 'eventually' takes, in generations, depends on the nature of the distribution, the pace of the response to climate change, and the amplitude of change. But the key component is appreciation of the role of an emergent property of species (distribution) as an element in the equation (Jablonski et al. 1985).

It is clear from the examples above that distributions of species are quite capable of extending across continents or oceans within a few thousands of years, which is effectively instantaneously on geological time-scales. It can be reasonably assumed that once a species has come into existence, it will occupy all accessible habitat within the continent or ocean of origin on this time-scale. This has an important practical consequence: it will normally be difficult or impossible to distinguish spread to a locality from speciation at the locality in a given fossil sequence, no matter how finely the record is sampled. In principle, localities with the earliest record of a new species should be nearest the origin of the species, but the required temporal resolution of less than thousands of years is only just within the possible resolution of the Holocene record, and well beyond the resolution of all earlier periods. For practical purposes, it will probably be best to take the purist stance that the origin of a species, by whatever mechanism, takes place at an instant in time and space that can never be observed. What the fossil record detects is population increase and expansion of the distribution of new species, and not their origin.

6 · *Biological response: evolution*

This chapter reviews evidence for evolutionary change in relation to the pace of climatic change. Most examples come from the late-Quaternary as the most accessible period of time for consideration of changes on Milankovitch time-scales, short by geological standards, but long ecologically. The chapter addresses directly the question that Lyell put to Darwin (see page 9), by considering whether or not the fossil record supports a notion of continual change, and hence that processes in operation today can be extrapolated back through all of time (substantative uniformitarianism), or a notion of disruption of ecological patterns on long time-scales, in which case no extrapolation is possible.

The evolutionary response of populations to climatic changes may, in principle, have taken place on a variety of time-scales. Populations might change gradually, through adaptation (Darwinian natural selection) or by any other cause, possibly to the extent that new species could be considered to have evolved (phyletic gradualism). Lineages might split to establish allopatric populations with sufficient new characteristics to form new species (cladogenesis). Populations (and species, when all constituent populations are lost) might become extinct. There may be no evolutionary change at all (stasis).

The quality of evidence needed to show that evolutionary change has taken place in response to climatic changes on Milankovitch time-scales is demanding. Ideally, time, place, and rate of change are needed from data with a temporal resolution of less than 10 kyr, and this is rarely available. Additionally, and even less frequently found, is evidence that observed change is indeed evolutionary, and not a consequence of movement of a stable geographic cline (Stuart 1982), nor the increase of population or expansion of distribution following a speciation event somewhere else (see Chapter 5). Some knowledge of what is happening all over the distribution of the species, and not just at the site of one fossil sequence, is needed to achieve this. Necessarily, these criteria mean focussing

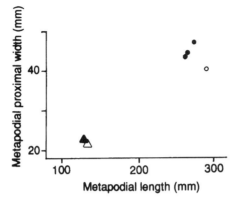

Figure 6.1 Comparison of *Cervus elaphus* (red deer) from the last interglacial at Belle Hogue, Jersey (triangles), and from Britain (circles). Filled symbols are metacarpels, open symbols are metatarsals. Redrawn from Lister (1989, Fig. 2d).

attention on Quaternary changes, since there is inadequate temporal resolution for any other part of the fossil record.

Geological time

Mammals

Many islands have, or had, populations of dwarf mammals with presumed ancestral forms on the adjacent mainland (Sondaar 1977). Usually it is not possible to date the time of arrival of the ancestral form on the island, and therefore it is also not possible to obtain estimates of the rate at which dwarfing took place. One instance where time and rate have been obtained is the dwarfing of red deer (*Cervus elaphus*) on the island of Jersey, about 25 km from the north coast of France (Lister 1989). Deposits on the island from the cold stage preceding the last interglacial (about 125 ka) show that red deer of the same size as contemporary French and British populations were then present, whereas last-interglacial red deer remains indicate a body size of only about one-sixth that of British red deer at the same date (Fig. 6.1, and other examples in Lister (1989)). A fall in sea-level of only 8 m would be enough to connect Jersey with the French mainland. Lister (1989) calculated from last-interglacial $\delta^{18}O$ ratios (which are due largely to the volume of continental ice, and hence related to sea-level) that sea-levels were high enough to isolate Jersey for only 9–10 kyr within the last interglacial. The available time for dwarfing is somewhat less than this (about 6 kyr) because of the geological setting

Table 6.1. *Stratigraphical ranges of deer during the British Quaternary*

	Br	P	Be	C	A	H	W	I	D	F
Eucladoceros falconeri	x									
Eucladoceros tetraceros		x								
Eucladoceros ctenoides		x								
Eucladoceros sedgwicki		x								
Megaceros verticornis				x						
Megaceros dawkinsi				x						
Megaceros savini				x						
Megaceros giganteus					x	x	o	x	x	
Cervus sp. (?*C. perolensis*)	x									
Cervus elaphus				x	x	x	x	x	x	x
Dama dama				x	o	x	o	x	o	o
Capreolus capreolus				x	o	x	o	x	o	x
Rangifer tarandus			x		x	o	x	o	x	o
Alces gallicus		x								
Alces latifrons			x							
Alces alces									x	x

Notes: Occurrences in the British Isles are marked by 'x', and occurrences elsewhere in Europe when not in the British Isles by 'o'. Abbreviations for stage names of interglacials: F, Flandrian (= Holocene); I, Ipswichian (= last interglacial); H, Hoxnian; C, Cromerian; P, Pastonian; Br, Bramertonian. Abbreviations for stage names for glacials: D, Devensian (last glacial); W, Wolstonian; A, Anglian; Be, Beestonian. Stages are presented in chronological sequence from oldest (left) to youngest (right).
Source: Compiled from Lister (1984a, 1986).

of the beach deposit containing the fossils (Lister 1989). Mainland European red deer had existed for the previous 400 kyr with only minor changes. Although there is no way of knowing whether the dwarfed island population was reproductively isolated from the mainland population (and hence speciation had taken place), this is clearly an example of rapid evolution in an allopatric isolate. The Jersey population did not persist later than the end of the last interglacial. Presumably, it became extinct after a fall in sea-level recombined island and mainland populations.

The record of deer faunas during the European (especially British) Quaternary has been reviewed by Lister (1984a, 1986). The stratigraphy of Quaternary deposits in Britain is reasonably well-known, and permits the placing of faunal remains in a stratigraphic context at the level of glacial–interglacial oscillations. Deer exhibit evolutionary and distribution changes within the Quaternary (Table 6.1). All living species have fossil records that extend back no further than the mid-Quaternary, and may thus be no more than about 600 kyr old (Lister 1986).

Two species of mammal in the northern hemisphere, the elk or moose (*Alces alces*) and the polar bear (*Ursus maritimus*) may be less than 100 kyr old, being known with certainty only from last-glacial and modern material in both North America (Kurtén & Anderson 1980) and Europe (Stuart 1982; Lister 1984b), and there are other species whose known stratigraphic records are confined to a single glacial–interglacial oscillation (Table 6.1).

Analysis by Barnosky (1990) of the third upper molars from fossil and modern populations of the Meadow vole (*Microtus pennsylvanicus*) in Virginia, USA, showed that these teeth have changed within the last 30 kyr in two morphometric traits, possibly triggered, indirectly, by the climatic changes between last-glacial and Holocene conditions. At the same time, four other traits remained statistically unchanged in Virginia, but three of them may have changed in Colorado populations (fossil confirmation is currently lacking). Barnosky (1990) suggested that intraspecific variation in *Microtus pennsylvanicus* develops in a mosaic pattern. Some changes persist locally through time, while others spread to more distant populations.

Fishes

The early Mesozoic sequences of annually-laminated lake sediments described by Olsen *et al.* (1982: and see page 78) have yielded a fish fauna that has been sampled in sufficient detail to estimate times for speciation in the genus *Semionotus* (McCune 1996). Twenty-one species from this genus occur in a single sedimentary unit from a single Early Jurassic lake of the Newark Basin, New Jersey, USA. Six of these species are considered to be endemics because they are not found in older deposits nor in deposits of the same age in other basins. The sedimentary units occur rhythmically at a periodicity estimated to be 21–24 kyr, enabling McCune (1996) to conclude that these six species originated within 5–8 kyr (the earliest third of the sedimentary unit), similar in time-scale to examples of rapid speciation in fishes of some late-Quaternary lake basins.

Beetles

Coope (1970) stated that neither he nor any of his colleagues had ever found a Quaternary fossil beetle intermediate in form between any

158 · Biological response: evolution

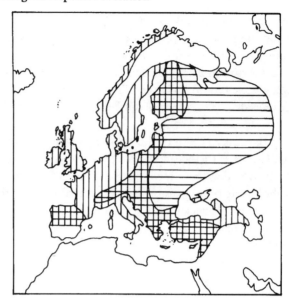

Figure 6.2 Modern distribution of *Helophorus aquaticus* (Hydrophilidae), an aquatic beetle, and its races. Vertical shading indicates distribution of the race with straight paramere margins, horizontal shading indicates distribution of the race with curved paramere margins. Cross hatch indicates the area of overlap and transition. Redrawn from Angus (1973, Fig. 53).

known living species, despite having examined many thousands of specimens from several dozen sites of middle and late-Quaternary age. Since then, many other sites and specimens have been examined, but with exactly the same conclusion: beetle species have maintained their integrity over, at least, much of the Quaternary, and thus over several to many climatic oscillations on Milankovitch time-scales (Ashworth 1979; Coope 1979, 1987, 1995). *Helophorus aquaticus*, an aquatic beetle, has been intensively studied by Angus (1973). It occurs across much of Europe and has two geographical races. The west European race has straight outer margins to the parameres of the male genitalia, the east European race has curved outer margins, and there is a narrow zone between the two where specimens have intermediate margins (Fig. 6.2). Quaternary fossil specimens have been found with the male genitalia intact and show clearly that the two races have maintained their integrity since about 120 ka (the last interglacial). However, the distributions of the two races have changed: the race with straight-margined parameres is found in Britain during Quaternary warm stages, and the race with curve-margined parameres occurred there during cold stages. Fossil remains of the intermediate form have been found in western Ukraine (Angus 1973; Coope 1979). It is clearly an oversimplification to view geographical variation as incipient speciation when a knowledge of past history of these races shows that both forms have maintained their integrity despite changes in distribution of both (Coope 1979). The only evidence of Quaternary evolution that Matthews (1974) could find among beetle remains in Alaskan early Quaternary sediments was a change in wing size of the flightless staphylinid beetle *Tachinus apterus*. This may be evidence for an intraspecific evolutionary change in beetles during the Quaternary, in which case it would be the only example known (Coope 1979).

Ostracodes

Analysis of morphological characters of late Cenozoic marine ostracodes from Atlantic, Caribbean, and Pacific coasts by Cronin (1985, 1987) shows long-term stasis during Quaternary climatic oscillations, but a period of rapid speciation at 3–4 Ma. Cronin (1985) analysed stasis by examining morphological variability of *Puriana mesacostalis* at a series of temporal scales through the late Cenozoic (Fig. 6.3), and found that variability increased only slightly as the temporal scale increased. Two species (*Puriana mesacostalis* and *Puriana floridana*) showed intraspecific variation over the past 3.5 Myr, but remained distinct, and no new species split from either during this time. But at about 3.5 Ma, at least six new species had appeared abruptly, in less than 330–500 kyr, from ancestral *Puriana rugipunctata* and *Puriana* aff. *elongorugata*, both of which persisted sympatrically with their descendant species for the next 3 Myr with discrete phenotypes. Cronin (1985) suggested that these speciations in *Puriana* took place because of unpredictable shifts in climatic and oceanographic conditions due to the formation of the Isthmus of Panama and consequent separation of Pacific and Atlantic oceanic circulations. The time, rate and place of the speciations are not identified by this study, but they do seem to have been sympatric. On the other hand, several other species whose ranges were split by the formation of the isthmus show no consequent morphological change when specimens from before and after the formation are compared (Cronin 1988; Cronin & Schmidt 1988).

Cronin (1987) described the appearance of three new species of *Puriana* along the east coast of North America and the Gulf of Mexico. Temperature-sensitive ostracodes moved north and south along this coast

160 · Biological response: evolution

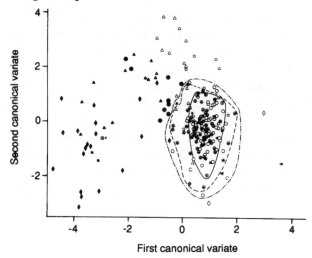

Figure 6.3 Morphological stasis in the ostracodes *Puriana mesacostalis* and *Puriana floridana*. Specimens of both species from late Cenozoic sediments of eastern North America are plotted on the first two canonical variate axes, accounting for 92.4% of the total variance. Filled symbols indicate specimens of *Puriana floridana*, open symbols *Puriana mesacostalis*. Age of specimens: squares, Pliocene; triangles, early Pleistocene; filled circles, middle Pleistocene (*Puriana floridana*); asterisks, middle Pleistocene (*Puriana mesacostalis*); open circles, late Pleistocene; circle with cross, Holocene; circle with dot, 1 ka; diamonds, Holocene. Three ovals outline levels of stratigraphical acuity for *Puriana mesacostalis*: inner, specimens from one sample (0.1–1 kyr); middle, specimens from one formation (10–50 kyr), and the outer, specimens from two formations (100–200 kyr). The larger ovals indicate greater morphological variability. The near concentricity and similar size of the ovals indicate morphological stability over 20 kyr of climatic change. Reproduced from Cronin (1985, Fig. 1B).

during glacial and interglacial climatic changes. He argued that *Puriana carolinensis* split off from ancestral *Puriana minuta* (which later became extinct) in a peripheral part of the ancestor's range. Other new species developed in sympatry. Pacific, Caribbean, and Central American species at various times included species, subspecies and rare short-lived species, which developed in isolation, separated from other populations by deep-water barriers, following successful rare dispersal events (Cronin 1987; Cronin & Schmidt 1988). There have also been populations of *Hermanites transoceanica* that have become established off remote islands and atolls in the Pacific and Caribbean, yet retained morphological stability for up to several million years despite isolation (Cronin 1988).

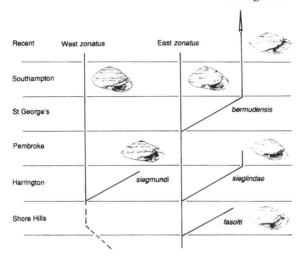

Figure 6.4 Reconstruction of the phyletic history of the Bermudan land-snail *Poecilozonites bermudensis* during stages of the Quaternary (named on left), showing iterative development of subspecies. Redrawn from Gould (1969, Fig. 20).

Molluscs

Investigations of the evolution of populations and species of land snails in and on Quaternary deposits of islands in the Caribbean and western North Atlantic Ocean come close to the ideal for the study of species history during the Quaternary. The entire geographical ranges of the species are small and tightly constrained, and specimens occur as fossils within the material on which they were living. In at least some cases, it is possible to detect the existence of new species with the confidence that they evolved at or near the site of deposition and cannot have moved in after an origin at some distant point.

Gould (1969) described the history of *Poecilozonites* (subgenus *Poecilozonites*), land-snails endemic to Bermuda, with a single extant representative, *Poecilozonites bermudensis*. The Quaternary geology of Bermuda consists of an alternating sequence of glacial red soils and interglacial carbonates. During glacials, lowered sea-level would have exposed an area of land ten times the size of modern Bermuda. During the last 300 kyr, the main stock of the modern taxon, *Poecilozonites bermudensis zonatus*, has branched at least four times (Fig. 6.4) and has also undergone morphological fluctuations correlated with glacial–interglacial oscillations. The branches are paedomorphic subspecies that arise at the periphery of

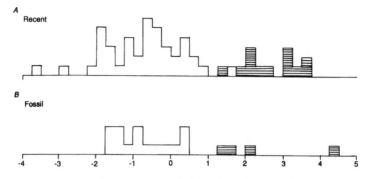

Figure 6.5 Histograms for projections of *A*, recent and *B*, fossil samples of the Bahamian land-snail *Cerion agassizi* from the islands of Eleuthera (open histogram) and Cat (shaded histogram) upon a discriminant axis. Both recent and fossil samples are distinguished in the same manner. Redrawn from Gould (1988, Fig. 13).

ancestral geographic ranges with loss of genes that characterize ancestral adult stages. Gould (1970) showed how land-snail size parameters of *Poecilozonites bermudensis*, *Thysanophora hypolepta*, and *Carychium bermudense* all vary in phase with Quaternary glacial–interglacial oscillations.

For New Providence Island, on Great Bahama Bank, Gould and Woodruff (1986) described finding five late-Quaternary species of the West Indian land-snail genus *Cerion*. *Cerion clenchi* occurs only in the oldest unit (>120 ka), *Cerion universum* occurs only in the 'middle' unit (120 ka), *Cerion agassizi* is found in the 'middle' unit and possibly also Holocene deposits (but is not found on the island today), *Cerion glans* is found in Holocene and modern deposits, and *Cerion gubernatorium* occurs on the island today, but has no fossil record on it. None of these species has an ancestor–descendant relationship with any of the others, so each must represent a new instance of spread to New Providence. *Cerion agassizi* is found in last-interglacial (120 ka) deposits and still persists on the islands of Cat and Eleuthera, the two nearest neighbour islands of New Providence, and connected to it by the last glacial fall of sea-level. Gould (1988) showed that populations on Cat and Eleuthera today have distinct sets of characters ('island signatures'), which have persisted since at least 120 ka, despite the opportunity for introgression between populations during the last glacial, and despite hybridization with *Cerion glans* during the Holocene (Fig. 6.5). Geographic variation in *Cerion agassizi* has thus been maintained for at least 120 kyr, despite the climatic changes of one

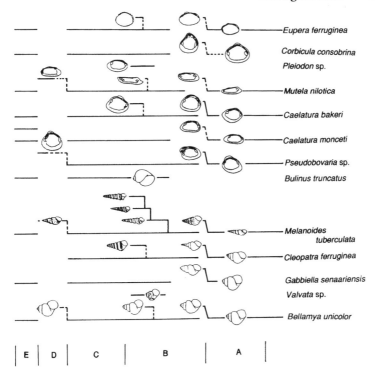

Figure 6.6 Summary of patterns of change in a Cenozoic sequence of molluscs from Turkana Basin, Kenya. Stratigraphic units (from oldest, on right, to youngest) are: A, Kubi Algi; B, Koobi Fora Formation, Lower Member; C, Koobi Fora Formation, Upper Member; D, Goumde; E, Galana Boi. One significant tuff horizon is the Suregei tuff at the upper boundary of Kubi Algi. Redrawn from Williamson (1981, Fig. 4).

full glacial–interglacial oscillation. Gould & Woodruff (1990) described how a fortuitous event (dispersal of a propagule of *Cerion dimidiatum* from Cuba to Great Inagua, Bahamas) produced a pattern of geographic variation there by hybridization with local populations of *Cerion columna*. This work illustrates not only the stability of variation in populations of these animals over long periods of time, but also how at least some of that variation can arise through factors that have nothing whatever to do with the local environment.

An important sequence of Cenozoic molluscs from the Turkana Basin, Kenya, is described by Williamson (1981). The fossil fauna was discovered within a 400-m thick accumulation of sediments that stretches laterally for 1000 km east of Lake Turkana. The molluscs are well-preserved and abundant in unconsolidated sediments, and they mostly belong to extant lineages. The sediments have been investigated extensively geologically, and there is an established chronostratigraphical context. Fig. 6.6 summarizes the patterns of evolutionary change seen in 13 lineages. A number of the lineages show simultaneous changes at the horizons of prominent tuff layers. Williamson (1981) argued that the changes reflect speciation events taking place within geographically-isolated populations, in accord with Mayr's (1954) model, and completed within 5–50 kyr. His interpretation was challenged by Fryer *et al.* (1983) on the grounds that the molluscs concerned were more variable than he had allowed for, and that contemporaneous changes in several lineages suggested some environmental control ('ecophenetic shifts'). In reply, Williamson (1985a) indicated that the morphological criteria he used were not subject to the high degree of variability mentioned by Fryer *et al.* (1983), and that the sediment accumulations were marked by evidence for many environmental changes, but morphological shifts are relatively rare, demonstrating an impressive degree of stasis through most of the time represented by the sediments. The subsequent debate, which continued through another pair of exchanges (Fryer *et al.* 1985; Williamson 1985b) is of interest chiefly for the way it casts light on the difficulties that can arise when there is more than possible interpretation of a dataset.

If Williamson's (1981) interpretation is correct, then we have here a series of lineages that have speciated a small number of times during a period of, at least, several hundred thousand years. The number of speciation events seen in any one lineage is much smaller than the number of orbitally-forced climatic oscillations that occured during this period of time.

Corals

Faunas of reef-building corals within the Indo-Pacific region are extremely homogeneous, with species that extend from eastern Polynesia to the western Indian Ocean (Potts 1984). Fossil records suggest that faunas have changed little since the late Tertiary, despite considerable sea-level fluctuations and consequent habitat shifts of these organisms (see Chapter 4).

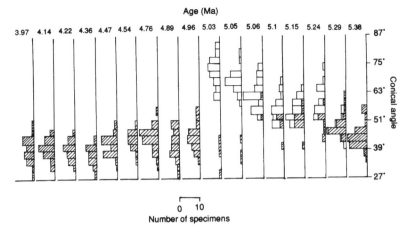

Figure 6.7 Speciation of the Tasman Sea foraminiferal clade *Globoconella*. Histograms show the conical angles of the *Globoconella pliozea* chronocline and the *Globoconella conomiozea terminalis* chronocline in the intervals before and after the speciation of *Globoconella pliozea*. Shaded portions represent *Globoconella pliozea* and its ancestor morphotypes, unshaded portions represent *Globoconella conomiozea terminalis*. Note the uneven temporal spacing of samples. Redrawn from Wei & Kennett (1988, Fig. 9).

Protista

Speciation of the foraminiferal clade *Globoconella* in the Tasman Sea is described by Wei & Kennett (1988). In the late Miocene (7–5 Ma), *Globoconella* populations formed a continuous morphological cline in several characters between 40°S and 26°S. At the beginning of the Pliocene, at the periphery of the distribution, the main stock (*Globoconella conomiozea terminalis*) disappeared and was replaced by a different form, *Globoconella pliozea*, within 10 kyr at 5.06 Ma (Fig. 6.7). Wei & Kennett (1988) considered this replacement to be a typical example of geographic speciation (*sensu* Mayr 1942). The new species maintained morphological stasis for the next 600 kyr. Wei & Kennett (1988) discussed reasons for thinking that the appearance of *Globoconella pliozea* represents geographic speciation rather than translocation of a geographic cline, but it should be noted that they do not discuss the possibility that *Globoconella pliozea* originated elsewhere, and spread within 10 kyr to the localities where they observed it. Simultaneously, the main stock evolved 'gradually' into a new chronospecies (*Globoconella sphericomiozea*) within about 170 kyr (5.14–4.97 Ma). Wei & Kennett (1988) argue that this pair of speciations

166 · Biological response: evolution

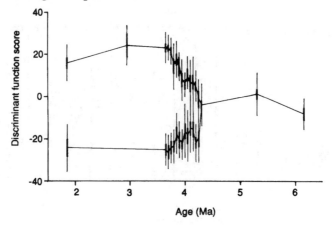

Figure 6.8 Evolution of equatorial Pacific radiolarians *Pterocanium prismatium* (lineage in upper left of figure) and *Pterocanium charybdeum* (lineage in lower left of figure and on the right), as shown by scores on a discriminant function separating the two lineages at 3.7 Ma. Plotted values are means ± 1 population standard deviation (thin lines) and ± 2 standard errors for the means (thick lines). Redrawn from Lazarus (1986, Fig. 11).

represent both punctuated equilibria and phyletic gradualism within the same lineage.

The equatorial Pacific record of the radiolarian species *Pterocanium charybdeum* and *Pterocanium prismatium* is described by Lazarus (1986). The former is extant today, but the latter originated from it at about 4.3 Ma, and then became extinct at about 1.8 Ma. The *Pterocanium prismatium* lineage evolved from *Pterocanium charybdeum* within 50 kyr (Fig. 6.8), but the two populations continued to diverge from each other over the next 500 kyr, accumulating about 10 standard deviations of difference (in multivariate indices of morphological difference) between them by about 3.5 Ma, with much slower rates of divergence until *Pterocanium prismatium* became extinct.

Sorhannus *et al.* (1988) described the evolution of the diatom *Rhizosolenia praebergonii* by lineage splitting from *Rhizosolenia bergonii* in the eastern and central equatorial Pacific at about 3.1 Ma. The first stage of development of *Rhizosolenia praebergonii* lasted about 400 kyr, at rates of evolution of 11–17 standard deviation units (in multivariate indices) every million years. The rate of change was most rapid early in the history of *Rhizosolenia praebergonii* (Fig. 6.9). During the second stage of its history, rates of change were more than an order of magnitude less.

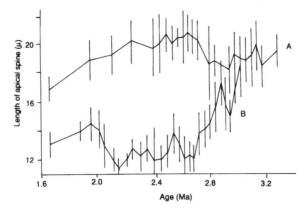

Figure 6.9 Morphological change through time in the marine Pacific diatom *Rhizosolenia* during the late Cenozoic. The lineage with longer apical spines (A) is *Rhizosolenia praebergonii*, and the lineage with shorter apical spines (B) is *Rhizosolenia bergonii*. Samples are connected at their mean values, and vertical bars indicate ± 1 standard deviation. Redrawn from Sorhannus *et al.* (1988, Fig. 3c).

Sorhannus *et al.* (1988) argued that this origination could not be an example of the punctuated equilibria model of evolution (Eldredge & Gould 1972) because it 'was not geologically instantaneous', defined as two standard deviation units between successive samples (sample interval averages 50 kyr in this work). Sample interval is an inadequate temporal basis for such a definition, being controlled by the researcher rather than being an independent measure of time. Sorhannus *et al.* (1988) attributed the evolution of *Rhizosolenia praebergonii* to selection towards a new adaptive peak, followed by stasis once the peak had been reached.

Plants

In western North America, pollen data are available demonstrating the spread northwards of lodgepole pine (*Pinus contorta*), thus giving the age at which populations became established at intervals northwards, and these have been related to the genetic structure and seed morphology of modern populations (Cwynar & MacDonald 1987). The comparison showed that overall genetic variability of these populations is not related to time since founding, but allelic diversity is significantly reduced by founding: younger populations have less diversity. Among the seed morphological parameters measured, wing loading (seed mass divided by the area of the seed wing) increases with time since founding. This means that populations founded most recently have seeds with lower wing loading, and hence are more dispersible. This result is perhaps not unexpected: individuals that produce seed which is more dispersible, other things being equal, are more likely to leave progeny at greater distances from themselves than individuals with typical seed. Northern populations of lodgepole pine, the most recently founded, are thus distinguishable from southern ancestral populations by allelic diversity and seed morphology, probably resulting from the series of founding effects needed to establish them. It has taken some 12 kyr, and dispersal across 2200 km to achieve this degree of population differentiation.

Bennett *et al.* (1991) have analysed the response of European forest trees to the end of interglacial periods, as well as to the beginnings. Tree distributions spread away from southern and eastern refugia at the beginnings of interglacials, extending, eventually across much of the continent. At the ends of interglacials, there is no return spread by ancestors of the relatively new northern populations. Instead, these northern populations become extinct *in situ* (Fig. 6.10). This has the important implication that any genetic changes that take place during an interglacial in the newly occupied ground are destined to be lost at the end of the interglacial when the northern populations become extinct as the climate cools towards the next glacial. Persistence of populations in southeastern areas during each interglacial may be just as important for their long-term survival in the Quaternary as persistence through range restrictions of cold stages. On the other hand, any genetic changes that take place in restricted cold-stage populations may become propagated across the continent as populations establish themselves successively northwards during interglacial periods.

Ecological time

The distribution and frequency of natural selection has been reviewed and discussed in a 'field guide' by Endler (1986). He identified studies demonstrating natural selection in 314 traits among 141 species, distributed widely among diverse plant and animal taxa. Demonstrations most commonly recognize mortality as the fitness component, and involve morphological traits, probably because these are easier to quantify than other aspects of fitness (for example fertility, mating ability for fitness, and traits involving physiology or biochemistry). Although Endler (1986) claimed that this is a remarkable number of demonstrations of

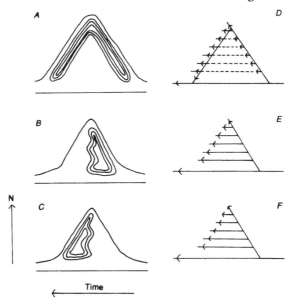

Figure 6.10 Diagrammatic representation of the spatial extent, variation in time, abundance, and ancestry of populations of of tree taxa in Europe during an interglacial.
A–C. Variation in northerly extent of range with time (bounded by outer lines), and abundance (indicated by contouring).
D–F. Ancestry of selected populations from *A–C*: populations become extinct at the end of the lines.
A and *D*. *Pinus* or *Betula*. Dashed lines indicate variability in persistence.
B and *E*. *Quercus* or *Ulmus*.
C and *F*. *Abies* or *Carpinus*.
Redrawn from Bennett et al. (1991, Fig. 4).

natural selection, it is interesting that there are so few, given the large number of species in the world ($2–20 \times 10^6$), and the alleged all-pervasive nature of the phenomenon. However, it is clear that natural selection is real, occurring throughout the living world, and it can be quantified and assessed by meticulous research.

Possibly the best known example of natural selection in the wild is also instructive because it involves strong environmental change. This is the classic case of Darwin's finches on the Galápagos. Darwin visited the Galápagos during the voyage of the *Beagle* in 1835. He collected a number of birds, and returned the skins to England, where they were examined by the ornithologist John Gould. Darwin was confused

Figure 6.11 Drawings of Darwin's finches (Fringillidae) of Galápagos and Cocos Island by W.P.C. Tenison, depicting a male and female of each species. 1, *Geospiza magnirostris*; 2, *Geospiza fortis*; 3, *Geospiza fuliginosa*; 4, *Geospiza difficilis*; 5, *Geospiza scandens*; 6, *Geospiza conirostris*; 7, *Camarhynchus crassirostris*; 8, *Camarhynchus psittacula*; 9, *Camarhynchus pauper*; 10, *Camarhynchus parvulus*; 11, *Camarhynchus pallidus*; 12, *Camarhynchus heliobates*; 13, *Certhidea olivacea*; 14, *Pinaroloxias inornata*. Reproduced from Lack (1947, Fig. 3).

by the birds, thinking they were a mixture of 'finches, wrens, "Grossbeaks", and "Icteruses"' (Sulloway 1982a, b, c; Desmond & Moore 1991). Gould (1837) realised that the birds were, in fact, all finches, and described them as an entirely new group of species. Since then, there have been a number of collections and taxonomic studies of the group (reviewed by Grant 1986), culminating in the modern judgement (Sibley & Monroe 1990) of 14 species, including one on Cocos Island (Fig. 6.11).

The Galápagos are a group of volcanic islands straddling the equator in the Pacific Ocean about 1000 km west of Ecuador. The climate is seasonal, with a hot, wet period from January to May, and a cooler and drier period for the rest of the year. However, this 'typical' pattern is not necessarily repeated every year, with especially large variation in the amount of rainfall. In some years there is heavy rain, associated with 'El Niño' events, from December through to May, but in other years there may be only a few light showers. The extent of rainfall determines the development of vegetation and production of flowers by individual plant species, which in turn affects the availability of seeds for consumption by birds. The 13 species of finch are distributed across the islands, usually with only a small number of species on any one island. They differ most strikingly in their beak sizes, and this turns out to be significant for determining which seeds can be tackled by which species (and by which individuals within a species). Lack (1947) suggested that a single species colonized the islands, at some (unknown) time in the past, and became differentiated into various forms following geographical isolation on the different islands of the archipelago. They became sufficiently different that, when populations met on the same island, they were already intersterile, and thus new species had originated. When the species met, there would have been competition, and two species could only coexist if each had become better adapted to one part of the food supply, or habitat, than the other. The present radiation thus resulted from a continuing process of ecological restriction and structural specialization. Since 1971, the ecological and evolutionary processes among Darwin's finches have been studied by Peter and Rosemary Grant and their colleagues (see, especially Grant (1986), Grant & Grant (1989), and a popular account by Weiner (1994)). Their work has been marked by the meticulous following of entire populations through several generations in the isolation of some of the smaller, undisturbed, Galápagos islands. It accounts for several of the demonstrations of natural selection in action included by Endler (1986).

During 1977 there was a severe drought in the Galápagos. On the island of Daphne Major, *Geospiza fortis* was subjected to strong directional selection (Fig. 6.12). Large birds with deep beaks survived best because they were better able to cope with the residual large, hard seeds that remained after the stock of small, softer seeds had been depleted. Female birds, smaller than males, suffered disproportionately, and in the following season the surviving females selected the larger from among the surviving males as mates, reinforcing the effect of the preceding natural

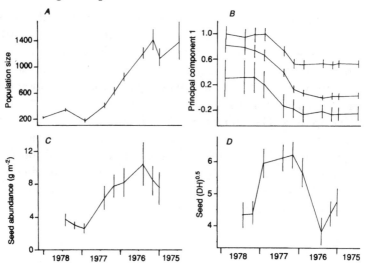

Figure 6.12 Temporal changes in numbers and morphology of the Galápagos finch *Geospiza fortis* (Fringillidae), seed abundance, and seed size on the island of Daphne Major.
A. Population estimates (means ± 2 standard deviations).
B. Principal component scores, as a measure of bird morphology (means ± standard error).
C. Estimates of seed abundance (means ± standard error).
D. Estimates of a measure of hardness (DH) for available, edible seeds (means ± standard error).
Redrawn from Boag & Grant (1981, Fig. 1).

selection. Thus, there was selection for larger birds with deeper beaks during drought years. However, small birds were selectively advantaged in two ways. They survived best in their first year, and small females started breeding earlier than their larger contemporaries. Grant (1986, p. 395) concluded, 'the selective shifts in opposite directions may roughly balance and be equivalent to a weak form of overall stabilizing selection'.

During 1983, there was an exceptionally long wet season. On the island of Genovesa, vines proliferated and smothered cacti, which died. This effectively eliminated cactus seed production, but increased the abundance of arthropods living in the rotting remains. During this period, adults of the large cactus finch (*Geospiza conirostris*) with deep and wide beaks were selectively advantaged, as they were capable of the tearing needed to extract arthropods from the dead cactus pads. By following examples such as these and their effects on characters across whole populations, Grant (1986) and Grant & Grant (1989) showed not only that selection is occurring, continuously, but also that it is varying continuously in strength and direction. In the case of these finches on Galápagos, the amount of rainfall in the 'wet' season is crucial for the development of vegetation and for the survival and breeding success of the finches. There are strong selection pressures, environmentally controlled, and high mortality and differential breeding success in consequence.

Discussion

For most fossil groups through most of the Quaternary, there is remarkably little evidence of evolutionary change. Work on groups as diverse as Quaternary beetles and plant remains is dominated by the same taxa over periods of time that are long relative to the periodicity of climatic change. Potts (1983, 1984), for example, argued that reef-building corals in the Indo-Pacific region exhibit little evolutionary change during the Quaternary, despite a growth zone that is never stable for more than a few thousand years, because their generation times, measured in decades, are long compared to other reef organisms. Corals may pass through 10–100 generations at a given level, while other reef organisms (fish, crustaceans, and molluscs) pass through thousands of generations and exhibit extensive Quaternary speciation. There are few known examples of evolutionary change, at any level, operating through the Holocene, and none where cause can be ascribed with confidence (and thus none that would meet the criteria of Endler (1986) to pass muster as 'demonstrations' of natural selection). Some work has been done on modern populations in relation to Holocene distribution changes, mostly to investigate relationships between populations as the process of spread continued.

Two of the architects of the modern synthesis noted geographic evidence that populations of a species may have diverged to at least subspecies level during the period since the last glacial, citing examples from, especially, vertebrates (Mayr 1942; Simpson 1944). Examples from other groups were added by Mayr (1963). These examples need treating with care, since they are based mostly on inference from modern distributions, and there is one example of the persistence of a beetle subspecies over more than one glacial–interglacial oscillation (see page 158), but the examples involving isolation are probably valid. Taken with the molluscan evidence described in detail by Gould (1970), there is clearly some speciation taking place within single oscillations of Milankovitch time-scales.

174 · Biological response: evolution

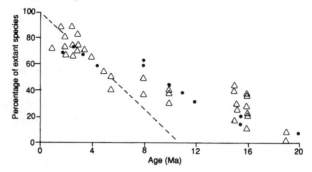

Figure 6.13 Lyellian curve for bivalve molluscs, produced from data representing the faunas of California (filled circles) and Japan (open triangles). The dashed diagonal line provides an estimate of mean species duration, and is approximately tangential to the curve at zero age. It's slope represents a mean species duration of 11 Myr. The number of species per sample varies between 12 and 150. Redrawn from Stanley (1985, Fig. 1).

Stanley (1979, 1985) presented data on mean species duration within groups by direct observation and by making estimates from curves of the percentage of species within fossil biota persisting to the present day (Lyellian percentages), plotted against the age of those biota (Fig. 6.13, Table 6.2). Both methods yield estimates for species durations of 1–30 Myr, depending on the taxonomic group (Table 6.2). Species are thus observed to persist much longer than the period of orbitally-forced climatic oscillations (Vrba 1985, 1993; Bartlein & Prentice 1989; Bennett 1990). This is true even for the current (Quaternary) series of oscillations with climatic changes enhanced to yield massive, continental-scale glaciation in parts of the northern hemisphere.

These long species durations are equivalent to about 10^5–10^7 generations, depending on group. Thus, as far as competent taxonomists can tell, there is so little net measurable morphological change over millions of generations that populations this far removed are considered identical at specific level (Stanley 1985).

At taxonomic categories above the species level, Buzas & Culver (1984) found that species durations on the Atlantic continental margin are shorter in shallow water (< 200 m) than deeper water, and also shorter in northern waters. Bottjer & Jablonski (1988) and Jablonski & Bottjer (1991) presented evidence that, among marine organisms, clades of shallow water have diversified more rapidly than those in deep water. In both cases, the observed phenomena are attributed to some aspect of the environ-

Table 6.2. *Estimates of mean species durations for a variety of biological groups, based on Lyellian percentages and direct observation of stratigraphic ranges*

Group	Estimated mean species duration (Myr)
Marine bivalves	11–14
Marine gastropods	10, 13.5
Benthic foraminifera	20–30
Planktonic foraminifera	>20
Freshwater fishes	3
Beetles	>2
Snakes	>2
Mammals	>1, ~2
Higher plants	>8
Bryophytes	≫20
Marine diatoms	25
Ammonites	~5 (mode of 1–2)
Trilobites	>1
Graptolites	2–3

Source: From Stanley (1985, Table 1).

ment. But it is also possible that speciation is taking place most rapidly in those environments most disturbed by changing sea-levels at frequencies on Milankovitch time-scales – namely, marine shallow water.

The examples discussed in this chapter show that species during the course of their history may exhibit three types of evolutionarily significant behaviour: stasis, gradual change, and speciation by lineage splitting. Stasis exists despite considerable environmental change, and is thus presumably dynamic behaviour (Gould & Eldredge 1993; Eldredge 1995). Morphology, at least, of a species can change with time. Speciation may take place by lineage splitting because environmental change enables establishment of new populations with a genetic make-up that is not representative of the organisms in the species originally. All these patterns of behaviour can and do occur. But stasis is the most frequent response to Quaternary climatic change, and the stability of species through these oscillations is impressive (Gould 1992). Natural selection has been shown to have occurred (for example, among populations of Darwin's finches), but there is no evidence that it accumulates over longer periods of time to bring about speciation in the Darwinian sense.

It appears that species are basically conservative, stable entities that can persist despite considerable environmental change. Occasionally, something unusual happens and speciation results, largely as a result of distribution changes resulting in turn from environmental change. Additionally, the characteristics of organisms in the species may change with time, regardless of environmental change, and presumably through the operation of processes unrelated to environmental change on Milankovitch time-scales.

Species may be sorted by geological and climatic changes with periodicities of 10^4–10^5 years (Milankovitch time-scales). This sorting is non-random, although the side effects may appear stochastic, and is also not heritable. Because sorting at one level of a hierarchy has effects that appear at all lower levels (Vrba & Gould 1986), it is tempting to say that the survival of a given species is due to adapted characteristics at organism level, but it might just as well be due to sorting for emergent characters of species, such as population size and distribution. It is, therefore, important to distinguish the two. Cracraft (1982) argued that geological changes brought about species sorting, but he was thinking of global scale, long-term (Earth history) lithospheric changes in complexity.

The life-spans of individuals are relevant here. Generation times vary greatly amongst different groups, from a few minutes for bacteria to a few years for many vertebrates, and up to a hundred years for trees. Thus, for most multicellular organisms, diurnal changes take place on a shorter time-scale than their lives, and can be considered as part of their environmental background. Many organisms live for much longer than one year, and for these, the annual cycle of seasons is a recurrent, predictable event. Typically, such organisms have behavioural and, or, physiological mechanisms that make possible a life that spans changing seasons. No organism, however, lives for periods of time that exceed the period of climatic oscillations, and so for no organism do these oscillations form part of their predictable environmental background. Just as an ordinary winter kills off most of the populations of insects that have life-spans shorter than one year and reproduce through several generations each summer, and just as mass extinctions remove a high proportion of taxa (at any level) at approximately 26 Myr intervals, so too oscillations of orbital changes have a profound influence on individuals and populations that persist typically on shorter time-scales.

On Galápagos, extreme weather events forced episodes of selection (see above). These events recur on time-scales of around the longevity of the longest lived individuals of Darwin's finches. But the periods of selection in one direction are followed by intervening periods of relaxed selection, or selection in a different direction. The net result is that populations of finches remain as recognizably the same species overall: there has been no speciation event within the period of study, and no species other than modern ones are known to have existed on Galápagos within the last few thousand years (Steadman 1986).

7 · Biological response: extinction

The third major type of response of organisms to global climatic change is extinction: failure to survive the new conditions. The Quaternary is characterized by its particularly extreme climatic oscillations, and also for the occurrence of wholescale extinctions of large mammals at about the glacial–Holocene transition (Grayson 1984a). Extinction has certainly occurred, but how widespread is it as a response to global climatic change? In the search for 'causes' or 'explanations' of the late-Quaternary mammalian extinctions, it is necessary to look away from mammals at this time and towards other groups at other times during the Quaternary. Is the transition at the beginning of the Holocene a particularly unusual transition compared to other parts of the Quaternary? There have been other oscillations of climate during the Quaternary as extreme as the transition at the beginning of the Holocene (for example Shackleton & Opdyke 1973; Martinson et al. 1987; see also Fig. 4.3). And why mammals, and not other groups?

Animals

Webb (1984) showed that there have been several periods during the last 10 Myr when rates of extinction among North American mammals have been high. The most severe episode, at about 5 Ma, involved 62 genera, of which 35 were large mammals (>5 kg body weight). The episode at the glacial–Holocene transition, by contrast, involved 43 genera, of which 39 were large mammals. There are three other episodes within the last 10 Myr with extinctions of a similar magnitude (Fig. 7.1). These high extinction periods appear to coincide with the ends of glacial phases (Webb 1984). Gingerich (1984) examined Cenozoic extinctions against the background pattern of originations, and found that, during the Quaternary and throughout the Cenozoic, rates of extinction and origination are closely related (Fig. 7.2), and that the rate of appearance during the last glacial stage exceeds the rate of extinction. However, taking the large

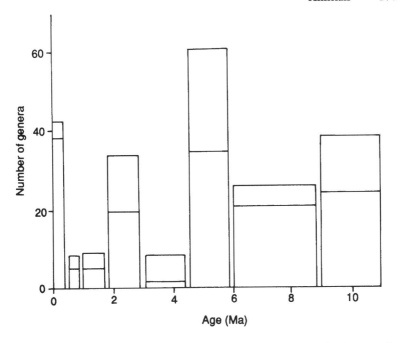

Figure 7.1 Late Cenozoic extinction episodes for North American land mammals. Bar heights indicate the number of extinctions within the period of time indicated on the age scale. The lower portion of each bar refers to genera of large mammals (body weights more than 5 kg), the upper portion refers to small mammals. Redrawn from Webb (1984, Fig. 9.2).

herbivorous mammals in isolation, there were more extinctions than originations. Gingerich (1984) argued that late-Quaternary extinctions must be seen as part of overall faunal turnover.

Maglio (1978) showed that rates of mammalian extinctions in Africa were higher in the Tertiary than in the Quaternary, and that within the Quaternary, rates of extinction were higher during the early-Quaternary than in the late-Quaternary. He, like Gingerich (1984), suggested that extinctions must be seen as part of high turnover. Mammals have evolved rapidly during the Cenozoic relative to other groups (Stanley 1979, 1985), and the high rate of generation of new species has been matched by high rates of extinction.

There is little evidence that groups other than mammals suffered unusually high extinction rates during the last glacial–Holocene transition. Coope (1995) emphasized the lack of extinction among beetle species during the Quaternary. There were extinctions of some species of birds

180 · Biological response: extinction

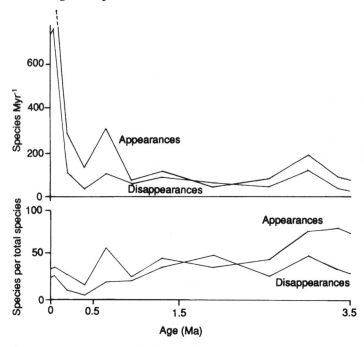

Figure 7.2 Patterns of change in the appearance and disappearance of all species of North American Plio–Pleistocene mammals. Rate of Wisconsin (since 0.1 Ma) appearances is 1030 Myr^{-1}. Note that age scale is not linear. Redrawn from Gingerich (1984, Fig. 10.3).

contemporaneously with the loss of mammals in North America, but these seem to have been nearly all scavenging or commensal species that were dependent on a large mammalian herbivore fauna. Their loss appears to have been a consequence of the loss of the large mammals, and lends no support to any hypothesis about the extinction of large mammals (Steadman & Martin 1984).

During the Holocene, several island groups have experienced considerable loss of vertebrate species that can be attributed to human colonization of those islands. Examples include the spectacular moas of New Zealand (Trotter & McCulloch 1984; Anderson 1989) and elephant birds of Madagascar (Dewar 1984), but recent excavations are revealing a rich fauna of extinct birds on most Pacific and Atlantic islands (for example Olson 1973; Olson & James 1982; Steadman *et al.* 1984; Steadman 1986, 1989; Pimm *et al.* 1995). Some Pacific island groups may have lost half of their bird species following colonization by Polynesians (Pimm *et al.*

1995). Madagascar lost a rich mammal fauna during the Holocene (Dewar 1984), and so did other remote islands (Sondaar 1977). The association of these Holocene extinctions (and also visible historic extinctions) with some aspect of human activity has been used to support the hypothesis that the last glacial extinctions of large mammals have also been brought about by human activity, or 'overkill' (Martin 1984). However, the acceptance of an anthropogenic explanation for extinctions in the late-Holocene does not necessarily mean that the same mechanism brought about extinctions at the last glacial–Holocene transition. Grayson (1984b) has argued that the overkill hypothesis is now so overextended that it is no longer falsifiable, and therefore no longer useful as an hypothesis explaining last glacial extinctions. He suggests that the role of the overkill hypothesis has effectively been taken over by climatic hypotheses about the same events, because these are falsifiable, whereas the overkill hypothesis is not. The degree to which people were responsible for late-Quaternary large mammal extinctions is still not resolved.

Plants

There have been extremely few recorded non-anthropogenic extinctions of plants during the Quaternary (Leopold 1967; Kershaw 1984). Many species have become locally extinct as a result of range changes, and some of these extinctions, especially those based on pollen data at generic level, may conceal a greater number of species level extinctions. For example, the tree genera *Carya*, *Eucommia*, *Tsuga*, and *Liquidambar* have occurred in Europe during the Quaternary, but no longer do so, and the tree flora of eastern Australia included the podocarps *Dacrydium*, *Phyllocladus*, and southern beech of the *Nothofagus brassii*-type group (see Chapter 5). In New Zealand, several genera are known to have become extinct locally since the late Cretaceous (Mildenhall 1980). These include Gondwanan relicts (such as the podocarps *Microcachrys* and *Microstrobos*) and also types that reached New Zealand well after it was isolated from other land masses but which have not persisted until the present (such as *Acacia* (early Pliocene to middle Quaternary), *Casuarina* (Paleocene to early Quaternary), and *Eucalyptus* (Miocene to early Quaternary)). Although these losses occur during the Quaternary, there is no obvious temporal patterning to such regional extinctions that might indicate why these plants became extinct at one time rather than any other.

A notable example of late-Quaternary plant extinction has been

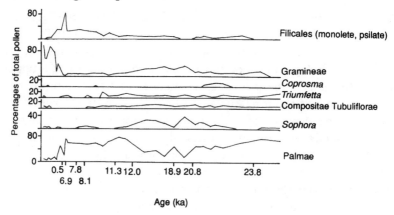

Figure 7.3 Abundances of selected pollen and spore types from the late-Quaternary pollen sequence at Rano Raraku, Easter Island. Redrawn from Flenley *et al.* (1991, Fig. 18).

described from Easter Island, in the southeastern Pacific (Flenley & King 1984; Flenley *et al.* 1991; Bahn & Flenley 1992). This is the most isolated fragment of presently inhabited land in the world, and formerly supported an extraordinary megalithic culture characterized by the building of giant statues of sculptured stone. The island is of volcanic origin and has three volcanic craters that are now lake-filled. Vegetation is predominantly grassland, and there are only two species of non-introduced tree, the endemic *Sophora toromiro* (Leguminosae) now extant only in botanic gardens, and *Triumfetta semitriloba* (Tiliaceae), and two shrubs, *Caesalpinia bonduc* (Leguminosae) and *Lycium carolinianum* (Solanaceae). Pollen analyses of sediments from the craters (Fig. 7.3), however, reveal, additionally, the former presence of a palm (Palmae) of an extinct species, related to the Chilean species *Jubaea chilensis*, a species of *Coprosma* (Rubiaceae), and one or more shrubs from the Compositae. Presence of the palm on the island is confirmed by the discovery of palm fruits in cave deposits, radiocarbon-dated to about 800 years ago (Dransfield *et al.* 1984). The pollen of all trees and shrubs decreased between about 1200 and 800 years ago, and deforestation was complete by about 500 years ago. It appears to have been the result of human activity because the decline happened entirely within the period that people have occupied the island, and the nature of the decline and associated events (evidence for burning, introduction of non-native species of plant and animal) are entirely consistent with human activity. Thus, the island formerly supported a more diverse flora of woody plants than it does today, and at least one of these, the palm, is now completely extinct, almost certainly the result of anthropogenic activity on the island. Several other woody plant taxa are now extinct on the island, but are not yet identified precisely enough to be sure whether they were members of species still extant elsewhere in Polynesia.

Discussion

The frequency of extinctions within the Quaternary in response to orbitally-forced climatic change is low, and clearly not a typical response. The recorded examples of extinctions within the Quaternary are few and far between. A high frequency of the extinctions that have occurred took place within the late-Quaternary, and especially the Holocene, and are probably or certainly brought about by anthropogenic activity. Most species survive most of the time, despite dramatic climatic oscillations.

Extinctions during the Holocene, of animals and plants, were probably typically caused by human activity, as today. But for earlier periods we are not yet in a position to say whether extinction is typically brought about by environmental change or by some other factor, such as competition. Species and organisms are sorted by climatic change, and those with characters that will not work under new circumstances do not survive. In the case of organisms, there are many characteristics that might bring this about, and this sorting can, carried to extreme, be upwardly causal: when all organisms of a species are dead, the species is extinct. Species may be sorted simply by distribution, for example: those with extensive distributions away from continental ice-sheets might survive, but those only near or on areas that will be covered by ice-sheets do not, even if both have similar characteristics emergent at organism level. Other characters emergent at species level that might enable sorting include population structure, and density. Such sorting is non-random, and would tend to generate lineages of species within certain emergent characteristics, notably with respect to distribution. The European tree flora, for example, comprises species whose distributions are typically either widespread or southern. Species with exclusively northern distributions do not survive glacial periods.

8 · *Evolution and ecology: synthesis*

The task of linking evolution with ecology is largely a question of appreciating the significance of a range of time-scales. Darwin (1859) effectively decoupled them by placing ecology as the process that controlled the evolution, and the two disciplines have not recovered. This chapter aims to redress the balance. Relevant processes are active at all time-scales. None is more important than any other, and all must be incorporated into any synthesis of the organization of life.

Biological responses

Species clearly respond to global climatic oscillations on Milankovitch time-scales by distribution change, evolution, and extinction. The evidence discussed in Chapters 5–7 indicates that distribution change is much the most frequent response, associated with local extinction. Evolutionary change in any form is rare, but examples are known, and these may be interpreted as gradualistic, punctuated, or equivocal, but given the overall low frequency of any evolutionary response, it must be concluded that, for most species, most of the time, stasis is the rule through climatic oscillations of Milankovitch time-scales (see also Williams 1992).

Post-modern evolutionary synthesis

The record of environmental change during the Quaternary is undoubtedly of great interest for the history of modern communities, and also for the evolution of a proportion (as yet unknown) of modern taxa. This history assumes a greater significance, however, if it can be read as a model for the way taxa have responded to climatic changes of Milankovitch time-scales throughout Earth history. We need to think of the record of the Quaternary as the expression of dynamism at time-scales of 10^4–10^5 years, and not as history. If we do this, then the significance of the record of Quaternary environmental change for future change should become clearer (Kareiva et al. 1993).

It appears that the climatic changes of the Quaternary are paced by astronomical forcing, so it is likely that the forcing has been a factor in pacing climatic change throughout Earth history. The deep-sea $\delta^{18}O$ record shows that climate has been changing with frequencies predicted from orbital forcing since at least 3 Ma, before the Quaternary ice-ages began, and geological data from earlier periods provide evidence that orbitally-forced climatic oscillations have, as expected, given their astronomical cause, been a factor in pacing climatic change on Earth throughout its history (see Chapter 5). The Quaternary has been a period of extensive continental glaciations, but this in itself is not an unusual feature of Earth history (see Table 4.1). It is therefore worth considering events during the Quaternary as representative of the 20–100 kyr-time-scale throughout the fossil record. The appearance of continental ice-sheets during the Quaternary has led to extreme climatic fluctuations, especially in parts of the northern hemisphere. Pre-Quaternary climatic changes forced by orbitally-forced climatic oscillations may have been of lower amplitude than Quaternary changes (Sergin 1979, 1980), but a globally-averaged change of even 1°C is enough to disrupt communities. Terrestrial and marine communities have responded in the same way to climatic changes on this scale during the Quaternary in the tropics as they did to the larger changes near the ice-sheets (Chapter 5). Thus, globally, species have responded individualistically to climatic changes forced by orbital variations in the Quaternary, and it is likely that they have responded in a similar way to pre-Quaternary climatic oscillations. We can expect that disruption of communities has been a permanent feature of 20–100-kyr time-scales, although usually unrecognized because of the relatively coarse resolution of most of the palaeontological record.

The palaeoecological data presented in Chapter 5 reveal, above all, the individualistic behaviour of taxa in response to climatic change. Communities break-up and re-form readily, across all taxonomic groups, in all areas of the Earth. They can, potentially, recur in time and space, as and when the necessary members have overlapping distributions, and so are not spatiotemporally-bounded. Communities can thus be defined only by their membership, and should be considered as classes, rather than as individuals. Because communities are transient, on time-scales much less than the longevity of the species that may belong to them, competitive interactions between populations are also transient, changing from one period to another as environmental change forces community reorganization. The processes of distribution change also bring about large fluctuations in population sizes. There are clearly implications for the importance of natural selection for evolution.

Endler (1986) assessed the importance of natural selection for evolution, identifying four views, termed, for brevity, 'selection', 'random', 'equilibrium', and 'constraint'.

(i) The selection view proposes that the effects of natural selection dominate other evolutionary factors, and are the most important factors in evolution. Mutation and recombination generate genetic variation, which then allows rapid response to changing environments.

(ii) The random view proposes that mutation, recombination, and genetic drift play the creative role in evolution, and that natural selection has a minor role. Most non-deleterious mutations or variants are selectively neutral, or nearly so, and only a small proportion are advantageous. There is not necessarily sufficient genetic variability for rapid response to changing environments, so populations either remain unchanged or become extinct, unless new mutations arise. This view appears to be more successful at explaining biochemical and molecular variation than morphological variation.

(iii) The equilibrium view is similar to the selection view, but starts with the assumption that natural populations are at an equilibrium, implying that environments have remained constant for long enough for this to have been achieved. It also assumes that there is enough genetic variation to reach any phenotype, and derives models that attempt to predict where the equilibrium should be depending upon trade-offs between internal pressures (physiology, biomechanics, and so on) and external pressures (predation, competition, and so on). This view grades into both the selection and constraint views.

(iv) The constraint view depends upon the idea that all mutations are not equally likely, but depend upon development and genetic systems, which in turn depend upon phylogenetic history, and the idea that genes and traits interact at various levels. There are, therefore, constraints on the sort of allelic substitutions that can take place, and hence on the rate and direction of evolution. Possible constraints include limits to variation determined by phylogeny, physiology, genetics, development, and so on, all inter-related, constraining natural selection, and evolution, possibly biasing it in certain directions. Natural selection can only affect changes in the frequency of variants once they have appeared.

Table 8.1. *Expanded deductive model of evolution*

IF potential exponential increase of populations	AND IF observed steady-state stability of populations	THEN struggle for existence = TRUE
IF organisms vary	AND IF struggle for existence = TRUE	THEN evolution by natural selection
	ELSE	THEN evolution in geographic isolation

Note: No temporal or spatial reference is included or implied because which of the two conclusions is reached depends upon the spatial and temporal scale: see the text for discussion.

Endler (1986) came to no conclusion about the validity or otherwise of these views: all have their merits and problems. He prefered to concentrate on distinguishing the origin of variation from natural selection, and emphasizing the different viewpoints of those who work on one or other aspect of the system. He argued that evolutionary studies would be considerably enriched by equal consideration of both origins and replacements. The palaeoecological record can, however, cast some light on the views he presented.

Environments are not constant: they change continually, driven on all time-scales by a variety of non-biological factors (plate tectonics, orbital variations, and so on). These in turn drive change in biological systems through abundance and distribution changes of taxa, and the biological interactions rapidly become immensely complex. It seems unlikely that equilibrium is attainable, on any time-scale. Equally, it seems unlikely that change is often continuous enough in any one direction to permit a constant selection pressure, particular with respect to competition. Additionally, as Lyell (1832, p. 174, quoted on page 8) pointed out, at any one place it is likely that neighbouring species would be able to immigrate following environmental change more rapidly than the original species could 'evolve'. We observe that the same taxa persist unmodified through major environmental changes, although natural selection can and does occur in the wild (Endler 1986).

Darwin (1859) presented his argument for natural selection deductively (see page 12), formalized by Huxley (1942) and Mayr (for example 1993). The argument rests upon the holding of certain conditions, and then a conclusion. But what if the conditions do not hold? Table 8.1 presents an expanded model of Darwin's argument. As climates fluctuate on Milankovitch time-scales, the tendency for populations to increase exponentially is realised, distributions increase enormously, and any struggle for existence is relaxed or eliminated (Elton 1927: and see page 36). But evolution can still occur in geographically-isolated populations. During periods of relative climatic stability, on ecological time-scales, evolution can proceed by natural selection. The model can thus be nested within a framework of different time-scales and spatial scales with an outcome appropriate for each scale.

The Quaternary record of environmental change and species response broadly supports the punctuated equilibria (Eldredge & Gould 1972) view of macroevolution. Stasis is the rule, despite dramatic climatic change. If gradual changes are taking place, continuously across fluctuating environmental conditions, then they are forced by some process other than natural selection, because selection pressures cannot be constantly in the same direction for long enough to achieve this. But the punctuated equilibria hypothesis is a statement about the observed patterns of the geological record, not a statement about process. Eldredge & Gould (1972) offer allopatric speciation (Mayr 1942) as the process of speciation, but stasis must also be an active process under changing environmental conditions forced by orbital variations on Milankovitch time-scales (Gould & Eldredge 1993).

So, how does life evolve? It appears that species are, potentially, long-lived, stable entities. They have distributions, and maybe subdivided into populations with greater or lesser degrees of interpopulation contact and interbreeding. Climates and environments have changed throughout Earth history on the same time-scales as during the Quaternary, and distributions shift in response, each species behaving individualistically. Some populations disappear, others increase rapidly, often at an exponential rate initially. Certain populations, perhaps typically towards the fringes of the original distribution, may become isolated. Other populations may be established as the result of dispersal. Which populations increase or decrease is effectively a random process depending on how the environmental changes fall with respect to the species' original distributions. Decrease and extinction may remove components of the gene pool from the species, while increase may allow the rapid expansion of small populations with characteristics that happen to be untypical of the population as a whole.

Repeated climatic change acts like the kneading of dough, squeezing populations down, then allowing expansion from small beginnings. The

Table 8.2. *Temporal hierarchy of dominant processes controlling evolutionary patterns seen in the geological record*

Tier	Periodicity	Cause	Evolutionary process
First	–	Natural selection	Microevolutionary change within species
Second	20–100 kyr	Orbital forcing	Disruption of communities, loss of accumulated change
Third	–	Isolation	Speciation
Fourth	~26 Myr	Mass extinctions	Sorting of species

Notes: The changes of the second and fourth tiers undo any evolutionary changes accumulated at the first and third tiers, respectively. The first and third tiers are defined temporally as intervals between events of the second and fourth tiers, respectively.
Source: From Bennett (1990, Table 1).

main part of the distribution may be largely unaffected, but peripheral populations will be continually created, and most will eventually be lost. But some will expand, and previously rare characteristics may become numerous. It is inevitable that speciation will be a feature of this situation, sooner or later. The opportunities for isolation are just too great for anything else.

The hierarchy of three tiers identified by Gould (1985: and see page 29) can be expanded to four tiers, and this provides a fuller understanding of the processes that control evolutionary patterns seen in the geological record, and a proper integration of palaeontological and ecological evidence (Table 8.2). Small adaptive changes may accumulate during ecological time (first tier), but most of these are lost as communities become re-sorted as a result of climatic changes of Milankovitch time-scales (20–100 kyr: second tier). Speciation takes place in geographically-isolated populations created by perpetual environmental changes of Milankovitch time-scales (third tier), but accumulated trends in lineages may be lost in mass extinctions (approximately 26 Myr: fourth tier). The first and third tiers are thus defined temporally as intervals between processes of the second and fourth tiers, respectively, but each tier is a distinct evolutionary process. Orbitally-forced climatic oscillations dislocate any interspecific interactions involved in the working of communities at the first tier by introducing disruption on a longer time-scale, thus serving to promote stasis. Accumulated adaptations under the previous relatively stable conditions within a single climatic oscillation are likely to be lost unless they also prove useful as 'preadaptations' (Simpson 1944) or 'exaptations'

(Gould & Vrba 1982) under the new conditions. Only rarely is it likely that adaptation could proceed in the same direction as before: the climate has changed, the species may be living in a different environment, and its competitors have changed. The processes involved in adjusting to change on the time-scales of orbitally-forced climatic oscillations must differ in kind as well as magnitude from the processes of life at the first tier. In this sense the effect of climatic oscillations has a parallel with mass extinctions (fourth tier), which introduce a different type of extinction pattern from the background rates of the third tier (Jablonski 1986).

The introduction of a tier between Gould's (1985) ecological moments and punctuated equilibria provides a mechanism that maintains the stasis of species by preventing any 'progress' (adaptive advantage) accumulating for more than 20–100 kyr. Without this process, progress would be able to continue up to the 1–10 Myr range before disruption by punctuated equilibria. If the punctuational model of Eldredge & Gould (1972) has any validity, there must be some such mechanism.

Most palaeontological research takes place at the time-scales of the third tier, which is too coarse to resolve what is happening at lower tiers. All ecological research takes place at the first tier, and the problem has been to integrate the two time-scales. Until recently, the Quaternary ice-ages seemed to form a barrier to this integration, because the one slice of time accessible to geologists for the intermediate time-scale seemed to be atypical for Earth history as a whole. Now it has been accepted that the ice-ages are expressions of perpetual variations in the Earth's orbit, it is clear that we can use the data of Quaternary research to illuminate processes in the history of life on time-scales between those of ecology and palaeontology: the type of change that has taken place in the Quaternary has been operating throughout Earth history. The result reinforces the concept of punctuational equilibria, and makes it more difficult to maintain the original thesis of Darwin (1859) that processes visible in ecological time gradually build up into the macroevolutionary trends seen in the palaeontological record.

Ecological research today is dominated, in the foreground, by the study of the interactions of populations, against a background of contribution to the study of the mechanisms of evolution. It assumes that natural selection is the mechanism of evolution. If there is any validity in the arguments and data presented in this book, then this is a false expectation. Interactions among and within populations may indeed lead to microevolutionary change, but there are processes operating on longer time-scales that combine to render unlikely the prospect that such changes contribute to evolutionary trends on a macroevolutionary scale. If all accumulated evolutionary change is wiped out by the next swing of climate, so is all the understanding of the ecologist who studied that change without consideration of its long-term significance.

Global climatic changes, recurring at Milankovitch time-scales, bring about considerable environmental change. Some aspects of this create conditions in which geographic isolation may occur: sea-level change, and changing extent of forest, for example. It has long been recognized that small populations may develop into new species (for example Mayr 1942), and climatic change can generate the reduction and isolation of populations (Stebbins 1984). Speciation is clearly not a typical response to climatic change in the Quaternary since species persist for periods of time much longer than the periodicity of environmental change, which is not the behaviour that would be expected from the Darwinian view or that of the modern synthesis. However, we need a mechanism to generate new species at time-scales much shorter than those of mass extinctions. If species are stable, spatiotemporally-bounded individuals, a mechanism is needed to force the shift or split from one species to another (or several). Climatic change can provide that force, but would be likely to do so rarely. While geographic isolation should occur as a result of global climatic change, there will only be a small proportion of species at any one time that have both small populations and which undergo speciation. Since climatic change is taking place on time-scales of 10^4–10^5 years, rare events at that time-scale become more likely as time-scales lengthen towards 10^7 years or longer. If there is a 1% chance that a species will speciate during any given global climatic oscillation, then it is near-certainty that it will have speciated after an order of 100 oscillations. This would generate species with durations in the observed range (10^6–10^7 years). Speciation has been observed in phase with glacial–interglacial oscillations (for example Gould 1970), and formation of subspecies has occurred within the period since the last glacial (see Chapter 6). Speciation on the same time-scales has been described from the early Jurassic (McCune 1996: and see page 157).

The mechanism for speciation presented by Eldredge & Gould (1972) is Mayr's (1942) allopatric mechanism. On geological time-scales, the process of shifting distributions, and speciation in isolated populations is likely to be invisible most of the time. The result may be speciation that looks as if it is sympatric, with one or more new species appearing within the range of the ancestral species, although old and new have undergone changes of distribution and abundance within a sampling interval. The process of divergence between ancestral and isolated populations might be accomplished in several ways, the chief requirement being simply that the process should be fast:

Sexual selection The process of finding partners by members of sexually-reproducing species may involve a choice by one sex or the other, and the consequences of that choice may lead to rapid development of phenotypic features. This process is termed 'sexual selection', and its existence and consequences were first explored by Darwin (1871), then taken up by Fisher (1930). They have been recently reviewed by Cronin (1991) and Ridley (1994). Where two characteristics exist, namely development of the feature in one sex (for example the tail of a male peacock) and preference for the feature when choosing a mate by the other sex, and where the choice confers a reproductive advantage:

> it is easy to see that the speed of development will be proportional to the development already attained, which will therefore increase with time exponentially, or in geometric progression. There is thus ... the potentiality of a runaway process, which, however small the beginnings from which it arose, must, unless checked, produce great effects, and in the later stages with great rapidity *(Fisher 1930, p. 136)*.

Fisher (1930) pointed out that the process would reach a period of stability, as checks (for example the feature in its extreme form may be sufficiently disadvantageous to counterbalance its reproductive advantage) came into play, and that the period of stability would then be far longer than the process of evolution of the feature by sexual selection. The process has been discussed most frequently among animals, but Willson & Burley (1983) have pointed out that there are also ways in which plants can exercise 'choices', with similar effect.

Polyploidy Genetically-related taxa may have chromosome numbers that are multiples of each other. Sometimes this has occured by simple doubling of the chromosomes (autopolyploidy), but it may also occur following hybridization (allopolyploidy). In either case, once the increase in chromosome number has taken place, the organism is immediately reproductively isolated from its diploid progenitors and their non-polyploid descendants. It thus provides a mechanism for speciation in a single generation and a mechanism for making sterile inter-specific hybrids fertile (Stebbins 1950). Polyploidy is frequent in plant groups, but much scarcer among animals. Details of the mechanisms of polyploidy and evolutionary consequences are described by Briggs & Walters (1984).

Post-modern evolutionary synthesis · 193

The mechanics of distribution change may have much to do with potential speciation. Extension of a distribution following a climatic change (or removal of a barrier, or whatever) requires organisms from the current species distribution. These organisms may be drawn from ancestors spread evenly across the species' distribution, or predominantly from ancestors at the range margin. Consider the case where dispersing organisms are not a genetically-representative sample of the species as then constituted. They establish beyond the original range margin, and their descendants extend the species distribution limits further out. Meanwhile, the organisms in the original distribution may or may not leave living descendants. When climate changes again, perhaps to something similar to original conditions, the newly occupied part of the distribution becomes untenable, the organisms occupying it do not leave ancestors locally, and the process repeats itself in reverse. If the species had continued to occupy the original distribution, the descendants of the original organisms would presumably be at a considerable numerical advantage over invading descendants from the new part of the distribution. If the invading descendants are, as a population, genetically-identical to the *in situ* descendants, then no invasion could be seen, even if it did take place. But the invaders are likely to be a non-random sample from a population that itself was established from a non-random sample of the original species. The probability that this type of back-and-forth shifting of distributions will, sooner or later, give rise to new species, through founder effects and genetic drift, must be high.

Although speciation taking place as described above is clearly allopatric, the allopatry arises through the *removal* of a barrier to dispersal and establishment by changing climate. The distribution has not crossed a barrier by some chance event, nor has a new barrier appeared to split the distribution. The allopatry arises where the new part of the species distribution consists of organisms descended from a small, non-representative set from those making up the species' original distribution. It is a consequence of the way distribution changes are related to the ancestry of the organisms concerned that the distribution changes amount to setting up allopatric populations of the species that may be of considerable size relative to the former distribution. Thus, the genetic make-up of the species may vary widely through time as a consequence of distribution changes.

In the only situation where there is any data on complete distribution movements over complete glacial–interglacial oscillations (trees in Europe), the evidence suggests that occupation of the new range is followed by local extinction, because environmental changes in the cooling phase of a glacial–interglacial oscillation are not the reverse of changes during a warming phase, particularly with respect to soil development (Bennett et al. 1991). In this situation, the representativity of the spreading part of the species (and its subsequent genetic history) is irrelevant for the long-term genetic history of the species, because it is all lost.

Species are spatiotemporally-bounded historical entities. The spatial aspect is controlled by climate. As climate changes, the distribution of a species is squeezed and pummelled into a new shape, and possibly into a new location. The rhythmic changes of climate of Milankovitch time-scales have the same effect on species as fists on a piece of dough. Most of the kneading involves changing shape and location of a single piece, but occasionally pieces of any size get squeezed off the main mass, or the mass splits. Thereafter, the pieces may rejoin seamlessly, or they may remain separate and from then on have separate histories as different historical individuals. Different species, with different locations and shapes of distributions react differently to climatic kneading. For some, low-frequency oscillations may be important, others may be more sensitive to high-frequency changes. Overlain on all this, are events of much lower frequency than Milankovitch time-scales (such as mass extinctions at 26 Myr intervals), or unique events (colliding continental plates, for example).

The Quaternary fossil record supports two long-held notions about communities from early twentieth century ecology. The wider significance of both was not appreciated at the time, and both were also ignored, or unseen, by palaeontologists. Gleason's (1926) individualistic concept of the plant association was published at a time when plant ecology was dominated by the concepts of 'climax', and the 'community as an organism', and was derided, until experiencing a rehabilitation 20 years later, by ecologists, and confirmation by Quaternary palaeoecologists another 20 years after that. Elton's (1927) rejection of the notion that populations were stable came at almost exactly the same time, but its significance was lost in a genetic and evolutionary world mesmerized by the developing theoretical population genetics of, for example, Fisher (1930) and Wright (1931). This, too, has eventually received support from Quaternary palaeoecology, albeit with populations fluctuating on time-scales much longer than those Elton was concerned with.

Vrba (1985, 1993) has developed an argument similar to that proposed here, but emphasizing the role of climatic change as a forcing factor for speciation. Her approach concentrated on identifying particular times when climatic change was sufficiently strong to generate synchronous speciation in a number of taxa: a 'turnover-pulse'. There needs to be emphasis on a continual background of speciation, at low, but variable, frequency. Certainly there must, at times, be more extreme episodes of climatic change, and consequently a higher number of speciation events. But over the long run of the history of life on Earth, we cannot necessarily expect to identify all those, and the most important point is to recognize that climatic change at Milankovitch time-scales can pace speciation, and hence evolutionary change. It decouples speciation from competition and other biotic interactions, and places greater control with aspects such as isolation and drift.

Difficulties

The ideas and hypotheses assembled here are not individually new, although the way that they are put together is. There is thus already literature on several of the key aspects, supporting and opposing. This section examines several objections that have already been raised to important parts of the argument presented here.

Recurrence of communities

On geological time-scales, a commonly-held view of communities is that they are 'recurrent'. Miller (1993) summarized various models of how recurrence is maintained, and contrasts the perceived recurrence of marine shelf communities with the non-recurrence of Quaternary terrestrial plant communities (see Chapter 5). The evidence of individualistic species response (and hence non-recurrence of communities) is over-whelming for those groups of organisms investigated in the mid- and late-Quaternary, so the issue is really one of whether recurrence occurs in other parts of the geological column. In part, the issue may be one of taxonomic resolution. In any given region, the same higher taxa (genera and families) may persist for tens of millions of years. Communities made of species drawn from this pool, at different times, may appear similar (and hence are 'recurring'), but do they have identical compositions at the species level? If species do respond to environmental changes individualistically, the chances of community recurrence must be greater for smaller taxonomic groups. As usual, the problem may be one of relative frequency of recurrence and non-recurrence, and of definition: how similar do two communities have to be before they can be said to be the same (and hence that a community has 'recurred')? Two communities comprise different individuals, so no two communities are ever identical.

On a philosophical level, Eldredge (1985) argued that communities do recur, so they might be considered as individuals. However, even if they did recur they would still not be individuals because they are defined by their membership, much as an atom of gold is defined by atomic number of 79, whenever and wherever it occurs.

Tracking the environment

It is frequently argued that species response to environmental change is such that species maintain their positions with respect to environmental gradients. Environmental change is thus not experienced directly, because descendant individual members of species have the same physical and biotic environment as their ancestors did before the change, but in a different geographical location. This is the phenomenon known as 'tracking the environment', and was invoked by Darwin (1859, p. 368, quoted on pages 13–14), to explain the lack of modification of species in response to climatic fluctuations of the glacial period, and by many others since (for example Coope 1979; Eldredge 1986, 1989).

To some extent, species clearly do, superficially, track the environment, but the 'fable does not bear close examination' (Williams 1992, p. 130). The issue is one of frequency and definition. How different do two environments have to be before we would reject the hypothesis that one species living in both really was experiencing different circumstances? For plants, at least, a shift of geographic location latitudinally alone brings about a changed environment because of their sensitivity to diurnal and annual light regimes. Spatially-varying climate and geology mean that no two areas are identical, and it is not realistic to expect that species really can 'track an environment' simply by shifting distribution at the same rate as some measure of climate. Dennett (1995, p. 294 no emphasis in original) commented that 'unless many species *moved in unison* in their habitat tracking, there couldn't be habitat tracking at all, since other species are such crucial elements in any species' selective environment'. But species do behave individualistically rather than in unison (Chapter 5), so communities break-up and re-form, changing biotic environments in the process. Climates are complex and multivariate, so it is unrealistic to imagine that global climatic change is expressed as a simple lateral translation of regional climates. On Milankovitch time-scales (10^4–10^5 years) there is sufficient happening environmentally in the way of changing climates, sea-levels, and extent of glaciation, that constancy of environment needs to be demonstrated by anyone seeking to argue that a species is 'tracking the environment'. The null hypothesis must be that all environments change constantly, and that species do experience changed environments over time without themselves being modified.

Distinctiveness of the Quaternary

It has been argued that the Quaternary is not typical of the rest of Earth history because of its extensive glaciation. Boucot (1990) identifies what he terms a 'Quaternary paradox': why is the pattern of biotic change as seen during the Quaternary different from that seen in the rest of geological time? He resolves it by arguing that the Quaternary record is a reflection of a 'high global climatic gradient, combined with the Milankovitch-cycle-generated, rapidly changing Quaternary environment that contrasts so markedly with the relatively unchanging Phanerozoic norm (when commonly sampled at million-year intervals or longer), plus the far more unchanging, buffered marine environment' (Boucot 1990, p. 555). Potts & Behrensmeyer (1992) discuss the response of organisms to climatic oscillations within the Quaternary, but question whether these responses are typical of the rest of the fossil record, hinting that the Quaternary may be a suitable model for those periods in Earth history that do have a glacial record, but not for the rest. In part this appears to result from the appearance of relative stability in pre-Quaternary periods, without consideration of the degree to which this is an artefact of the general coarsening of temporal resolution with increasing distance in time back from the present. Valentine & Jablonski (1993) acknowledge the importance of climatic oscillations for driving biotic change during the Quaternary, and the existence of such oscillations throughout the Phanerozoic, but doubt that biotic responses have approached the magnitude of Quaternary events, except during other periods of extensive glaciation, such as the Ordovician and Carboniferous.

The distinctiveness or otherwise of the Quaternary is clearly a crucial point for the whole of the argument presented here. However, it appears that the magnitude of climatic changes forced by orbital variations do not have to be particularly great to force substantial biotic responses. Chapter 5 describes numerous examples of biotic change in low latitudes that appear to have been forced by temperature changes of only a few degrees Centigrade, or by a shift of a monsoon belt. These types of climatic change are well within the scale of variation of climate expected on

198 · Evolution and ecology: synthesis

Milankovitch time-scales in parts of Earth history with minimal glaciation, such as the Cretaceous (see Chapter 4). The arguments in this book are based on these low-latitude changes, although the much greater mid- and high-latitude responses are cited as impressive (and exceptionally well-recorded) examples of what can happen.

The alternative is to ignore the Quaternary, and that leaves us with the unattractive prospect of essentially no data for a particular time-scale in Earth history.

References

Figures in square brackets are the page numbers on which each reference is cited.

Adam, D.P. (1988). Palynology of two Upper Quaternary cores from Clear Lake, Lake County, California. *United States Geological Survey Professional Paper* **1363**, 86pp. [109, 110]

Agassiz, L. (1840). *Études sur les Glaciers*. Neuchatel: Jent & Gasmann. [39, 40]

Alessio, M., Allegri, L., Bella, F., Calderoni, G., Cortesi, C., Dai Pra, G., de Rita, D., Esu, D., Follieri, M., Improta, S., Magri, D., Narcisi, B., Petrone, V. & Sadori, L. (1986). ^{14}C dating, geochemical features, faunistic and pollen analyses of the uppermost 10 m core from Valle di Castiglione (Rome, Italy). *Geologica Romana* **25**, 287–308. [96]

Allee, W.C., Emerson, A.E., Park, O., Park, T. & Schmidt, K.P. (1949). *Principles of Animal Ecology*. Philadelphia, PA: W. B. Saunders. [36]

Anderson, A. (1989). *Prodigious Birds: Moas and Moa-hunting in Prehistoric New Zealand*. Cambridge: Cambridge University Press. [180]

Anderson, A. & McGlone, M. (1992). Living on the edge — prehistoric land and people in New Zealand. In J. Dodson (Ed.), *The Naive Lands: Prehistory and Environmental Change in Australia and the South-west Pacific*, pp. 199–241. Melbourne: Longman Cheshire. [135]

Anderson, E.J. & Goodwin, P.W. (1990). The significance of metre-scale allocycles in the quest for a fundamental stratigraphic unit. *Journal of the Geological Society, London* **147**, 507–518. [89, 90]

Anderson, P.M., Bartlein, P.J., Brubaker, L.B., Gajewski, K. & Ritchie, J.C. (1989). Modern analogues of late-Quaternary pollen spectra from the western interior of North America. *Journal of Biogeography* **16**, 573–596. [115, 116]

Anderson, R.Y. (1982). A long geoclimatic record from the Permian. *Journal of Geophysical Research* **87C**, 7285–7294. [82]

Anderson, R.Y. (1984). Orbital forcing of evaporite sedimentation. In A.L. Berger, J. Imbrie, J. Hays, G. Kukla & B. Saltzman (Eds.), *Milankovitch and Climate, Part 1*, pp. 147–162. New York: D. Reidel. [82, 84]

Angus, R.B. (1973). Pleistocene *Helophorus* (Coleoptera, Hydrophilidae) from Borislav and Starunia in the western Ukraine, with a reinterpretation of M. Łomnicki's species, description of a new Siberian species, and comparison with British Weichselian faunas. *Philosophical Transactions of the Royal Society of London Series B* **265**, 299–326. [138, 158, 159]

Ashworth, A.C. (1979). Quaternary Coleoptera studies in North America: past and present. In T.L. Erwin, G.E. Ball & D.R. Whitehead (Eds.), *Carabid Beetles:*

their Evolution, Natural History, and Classification, pp. 395–406. The Hague: W. Junk. [158]

Ashworth, A.C., Schwert, D.P., Watts, W.A. & Wright, H.E., Jr (1981). Plant and insect fossils at Norwood in south-central Minnesota: a record of late-glacial succession. *Quaternary Research* **16**, 66–79. [138, 140]

Bahn, P. & Flenley, J. (1992). *Easter Island, Earth Island*. London: Thames and Hudson. [182]

Barnosky, A.D. (1990). Evolution of dental traits since the latest Pleistocene in meadow voles (*Microtus pennsylvanicus*) from Virginia. *Paleobiology* **16**, 370–383. [157]

Barnosky, C.W. (1984). Late Miocene vegetational development and climatic variation inferred from a pollen record in northwest Wyoming. *Science* **223**, 49–51. [73, 74, 75]

Barron, E.J. (1984). Climatic implications of the variable obliquity explanation of Cretaceous–Paleogene high-latitude floras. *Geology* **12**, 595–598. [61]

Barron, E.J., Arthur, M.A. & Kauffman, E.G. (1985). Cretaceous rhythmic bedding sequences: a plausible link between orbital variations and climate. *Earth and Planetary Science Letters* **72**, 327–340. [74, 76]

Barron, E.J. & Washington, W.M. (1984). The role of geographic variables in explaining paleoclimates: results from Cretaceous climate model sensitivity studies. *Journal of Geophysical Research* **89D**, 1267–1279. [61, 62]

Bartlein, P.J. & Prentice, I.C. (1989). Orbital variations, climate and palaeoecology. *Trends in Ecology and Evolution* **4**, 195–199. [174]

Beddall, B.G. (1988). Darwin and divergence: the Wallace connection. *Journal of the History of Biology* **21**, 1–68. [10]

Bennett, K.D. (1983). Postglacial population expansion of forest trees in Norfolk, UK. *Nature* **303**, 164–167. [151]

Bennett, K.D. (1986a). Competitive interactions among forest tree populations in Norfolk, England, during the last 10,000 years. *New Phytologist* **103**, 603–620. [151]

Bennett, K.D. (1986b). The rate of spread and population increase of forest trees during the postglacial. *Philosophical Transactions of the Royal Society of London Series B* **314**, 523–531. [115, 150]

Bennett, K.D. (1988). Holocene geographic spread and population expansion of *Fagus grandifolia* in Ontario, Canada. *Journal of Ecology* **76**, 547–557. [151]

Bennett, K.D. (1990). Milankovitch cycles and their effects on species in ecological and evolutionary time. *Paleobiology* **16**, 11–21. [174, 189]

Bennett, K.D. & Birks, H.J.B. (1990). Postglacial history of alder (*Alnus glutinosa* (L.) Gaertn.) in the British Isles. *Journal of Quaternary Science* **5**, 123–133. [107]

Bennett, K.D., Tzedakis, P.C. & Willis, K.J. (1991). Quaternary refugia of north European trees. *Journal of Biogeography* **18**, 103–115. [100, 105, 168, 169, 194]

Berger, A.L. (1976). Obliquity and precession for the last 5 000 000 years. *Astronomy and Astrophysics* **51**, 127–135. [46, 125]

Berger, A. (1978a). Long-term variations of caloric insolation resulting from the earth's orbital elements. *Quaternary Research* **9**, 139–167. [45, 50, 51, 62, 96]

Berger, A.L. (1978b). Long-term variations of daily insolation and Quaternary climatic changes. *Journal of the Atmospheric Sciences* **35**, 2362–2367. [45]

Berger, A. (1984). Accuracy and frequency stability of the Earth's orbital elements during the Quaternary. In A.L. Berger, J. Imbrie, J. Hays, G. Kukla & B. Saltzman (Eds.), *Milankovitch and Climate, Part 1*, pp. 3–39. New York: D. Reidel. [45]

Berger, A. (1988). Milankovitch theory and climate. *Reviews of Geophysics* 26, 624–657. [44, 45]

Berger, A. (1989). The spectral characteristics of pre-Quaternary climatic records, an example of the relationship between the astronomical theory and geosciences. In A. Berger, S. Schneider & J.C. Duplessy (Eds.), *Climate and Geo-sciences: a Challenge for Science and Society in the 21st Century*, pp. 47–76. Dordrecht: Kluwer Academic. [44]

Berger, A., Loutre, M.F. & Dehant, V. (1989a). Influence of the changing lunar orbit in the astronomical frequencies of pre-Quaternary insolation patterns. *Paleoceanography* 4, 555–564. [46, 47, 84]

Berger, A., Loutre, M.F. & Dehant, V. (1989b). Pre-Quaternary Milankovitch frequencies. *Nature* 342, 133. [46, 84]

Berger, A. & Pestiaux, P. (1984). Accuracy and stability of the Quaternary terrestrial insolation. In A.L. Berger, J. Imbrie, J. Hays, G. Kukla & B. Saltzman (Eds.), *Milankovitch and Climate, Part 1*, pp. 83–111. New York: D. Reidel. [45]

Betancourt, J.L. (1990). Late Quaternary biogeography of the Colorado plateau. In J.L. Betancourt, T.R. van Devender & P.S. Martin (Eds.), *Packrat Middens: the Last 40,000 Years of Biotic Change*, pp. 259–292. Tucson, AZ: University of Arizona Press. [119, 122]

Betancourt, J.L., van Devender, T.R. & Martin, P.S. (1990). Introduction. In J.L. Betancourt, T.R. van Devender & P.S. Martin (Eds.), *Packrat Middens: the Last 40,000 Years of Biotic Change*, pp. 2–11. Tucson, AZ: University of Arizona Press. [115]

Birks, H.J.B. (1981). The use of pollen analysis in the reconstruction of past climates: a review. In T.M.L. Wigley, M.J. Ingram & G. Farmer (Eds.), *Climate and History: Studies in Past Climates and Their Impact on Man*, pp. 111–138. Cambridge: Cambridge University Press. [42]

Birks, H.J.B. (1989). Holocene isochrone maps and patterns of tree-spreading in the British Isles. *Journal of Biogeography* 16, 503–540. [103, 109]

Birks, H.J.B. & Birks, H.H. (1980). *Quaternary Palaeoecology*. London: Arnold. [95]

Bloemendal, J. & deMenocal, P. (1989). Evidence for a change in the periodicity of tropical climate cycles at 2.4 Myr from whole-core magnetic susceptibility measurements. *Nature* 342, 897–900. [71]

Boag, P.T. & Grant, P.R. (1981). Intense natural selection in a population of Darwin's finches (Geospizinae) in the Galápagos. *Science* 214, 82–85. [172]

Bonnefille, R. & Riollet, G. (1988). The Kashiru pollen sequence (Burundi) palaeoclimatic implications for the last 40,000 yr B.P. in tropical Africa. *Quaternary Research* 30, 19–35. [126, 128]

Bottjer, D.J. & Jablonski, D. (1988). Paleoenvironmental patterns in the evolution of post-Paleozoic benthic marine invertebrates. *Palaios* 3, 540–560. [174]

Boucot, A.J. (1990). *Evolutionary Paleobiology of Behaviour and Coevolution*. Amsterdam: Elsevier. [197]

Bradley, R.S. (1985). *Quaternary Paleoclimatology: Methods of Paleoclimatic Reconstruction*. Boston, MA: Allen and Unwin. [48, 49]

Bradshaw, A.D. (1984). The importance of evolutionary ideas in ecology — and *vice versa*. In B. Shorrocks (Ed.), *Evolutionary Ecology*, pp. 1–25. Oxford: Blackwell Scientific Publications. [37, 38, 39]

Briggs, D. & Walters, S.M. (1984). *Plant Variation and Evolution* (Second edn). Cambridge: Cambridge University Press. [192]

Brubaker, L.B. (1975). Postglacial forest patterns associated with till and outwash in northcentral upper Michigan. *Quaternary Research* **5**, 499–527. [42]

Burkhardt, F. & Smith, S. (1990). *The Correspondence of Charles Darwin. Vol. 6 1856–1857*. Cambridge: Cambridge University Press. [9]

Bush, M.B. & Colinvaux, P.A. (1990). A pollen record of a complete glacial cycle from lowland Panama. *Journal of Vegetation Science* **1**, 105–118. [122, 123]

Bush, M.B., Piperno, D.R., Colinvaux, P.A., de Oliveira, P.E., Krissek, L.A., Miller, M.C. & Rowe, W.E. (1992). A 14300-yr paleoecological profile of a lowland tropical lake in Panama. *Ecological Monographs* **62**, 251–275. [123]

Buzas, M.A. & Culver, S.J. (1984). Species duration and evolution: benthic foraminifera on the Atlantic continental margin of North America. *Science* **225**, 829–830. [174]

Carozzi, A.V. (1967). *Studies on Glaciers Preceded by the Discourse of Neuchâtel by Louis Agassiz*. New York: Hafner. [40]

Carroll, R.L. (1988). *Vertebrate Paleontology and Evolution*. New York: W. H. Freeman. [5]

Chappell, J. & Shackleton, N.J. (1986). Oxygen isotopes and sea level. *Nature* **324**, 137–140. [67]

Chen, Y. (1988). Early Holocene population expansion of some rainforest trees at Lake Barrine basin, Queensland. *Australian Journal of Ecology* **13**, 225–233. [151]

Cockburn, A. (1991). *An Introduction to Evolutionary Ecology*. Oxford: Blackwell Scientific Publications. [29, 35]

COHMAP members (1988). Climatic changes of the last 18,000 years: observations and model simulations. *Science* **241**, 1043–1052. [51, 130]

Cole, K. (1985). Rate of change, species richness, and a model of vegetational inertia in the Grand Canyon, Arizona. *American Naturalist* **125**, 289–303. [118, 119, 120, 121]

Cole, K.L. (1990). Late Quaternary vegetation gradients through the Grand Canyon. In J.L. Betancourt, T.R. van Devender & P.S. Martin (Eds.), *Packrat Middens: the Last 40,000 Years of Biotic Change*, pp. 240–258. Tucson, AZ: University of Arizona Press. [118, 119, 120]

Collins, J.P. (1986). *Evolutionary ecology* and the use of natural selection in ecological theory. *Journal of the History of Biology* **19**, 257–288. [38]

Coope, G.R. (1970). Interpretations of Quaternary insect fossils. *Annual Review of Entomology* **15**, 97–120. [137, 157]

Coope, G.R. (1973). Tibetan species of dung beetle from Late Pleistocene deposits in England. *Nature* **245**, 335–336. [138]

Coope, G.R. (1977). Fossil coleopteran assemblages as sensitive indicators of climatic changes during the Devensian (last) cold stage. *Philosophical Transactions of the Royal Society of London Series B* **280**, 313–340. [138]

Coope, G.R. (1979). Late Cenozoic fossil Coleoptera: evolution, biogeography, and ecology. *Annual Review of Ecology and Systematics* **10**, 247–267. [41, 137, 138, 158, 159, 196]

Coope, G.R. (1986). Coleoptera analysis. In B.E. Berglund (Ed.), *Handbook of Holocene Palaeoecology and Palaeohydrology*, pp. 703–713. Chichester: Wiley. [137]

Coope, G.R. (1987). The response of late Quaternary insect communities to sudden climatic changes. In J.H.R. Gee & P.S. Giller (Eds.), *Organization of Communities: Past and Present*, pp. 421–438. Oxford: Blackwell Scientific Publications. [139, 158]

Coope, G.R. (1995). Insect faunas in ice age environments: why so little extinction? In J.H. Lawton & R.M. May (Eds.), *Extinction Rates*, pp. 55–74. Oxford: Oxford University Press. [137, 158, 179]

Coope, G.R. & Angus, R.B. (1975). An ecological study of a temperate interlude in the middle of the last glaciation, based on fossil Coleoptera from Isleworth, Middlesex. *Journal of Animal Ecology* **44**, 365–391. [138, 139]

Cracraft, J. (1982). A nonequilibrium theory for the rate-control of speciation and extinction and the origin of macroevolutionary patterns. *Systematic Zoology* **31**, 348–365. [176]

Croll, J. (1864). On the physical causes of the changes of climate during geological epochs. *The London, Edinburgh, and Dublin Philosophical Magazine and Journal of Science Fourth Series* **28**, 121–137. [10, 14]

Croll, J. (1865). On the physical cause of the submergence of the land during the glacial epoch. *The Reader* **6**, 435–436. [10, 14]

Croll, J. (1866). On the excentricity of the Earth's orbit. *The London, Edinburgh, and Dublin Philosophical Magazine and Journal of Science Fourth Series* **31**, 26–28. [10, 14]

Croll, J. (1867a). On the excentricity of the Earth's orbit, and its physical relations to the glacial growth. *The London, Edinburgh, and Dublin Philosophical Magazine and Journal of Science Fourth Series* **33**, 119–131. [10, 14]

Croll, J. (1867b). On the change in the obliquity of the ecliptic, its influence on the climate of the polar regions and on the level of the sea. *The London, Edinburgh, and Dublin Philosophical Magazine and Journal of Science Fourth Series* **33**, 426–445. [10, 14]

Croll, J. (1868). On geological time, and the probable date of the Glacial and the Upper Miocene Period. *The London, Edinburgh, and Dublin Philosophical Magazine and Journal of Science Fourth Series* **35**, 363–384. [10, 14, 41]

Croll, J. (1875). *Climate and Time in their Geological Relations: a Theory of Secular Changes of the Earth's Climate*. London: Daldy, Isbister. [10, 14]

Cronin, H. (1991). *The Ant and the Peacock: Altruism and Sexual Selection from Darwin to Today*. Cambridge: Cambridge University Press. [192]

Cronin, T.M. (1985). Speciation and stasis in marine Ostracoda: climatic modulation of evolution. *Science* **227**, 60–63. [159, 160]

Cronin, T.M. (1987). Evolution, biogeography, and systematics of *Puriana*: evolution and speciation in Ostracoda, III. *Paleontological Society Memoir* **21**, 71pp. [159, 160]

Cronin, T.M. (1988). Geographical isolation in marine species: evolution and speciation in Ostracoda, I. In T. Hanai, N. Ikeya & K. Ishizaki (Eds.), *Evolu-

tionary Biology of Ostracoda: its Fundamentals and Applications, pp. 871–889. Tokyo: Kodansha. [159, 160]

Cronin, T.M. & Schmidt, N. (1988). Evolution and biogeography of *Orionina* in the Atlantic, Pacific and Caribbean. In T. Hanai, N. Ikeya & K. Ishizaki (Eds.), *Evolutionary Biology of Ostracoda: its Fundamentals and Applications*, pp. 927–938. Tokyo: Kodansha. [159, 160]

Crowley, T.J. (1994). Potential reconciliation of Devils Hole and deep-sea Pleistocene chronologies. *Paleoceanography* **9**, 1–5. [66]

Crowley, T.J., Hyde, W.T. & Short, D.A. (1989). Seasonal cycle variations on the supercontinent of Pangaea. *Geology* **17**, 457–460. [56]

Cwynar, L.C. & MacDonald, G.M. (1987). Geographic variation of lodgepole pine in relation to its population history. *American Naturalist* **129**, 463–469. [167]

Dalrymple, G.B. (1991). *The Age of the Earth*. Stanford, CA: Stanford University Press. [40, 41]

Darwin, C. (1859). *On the Origin of Species by Means of Natural Selection, or the Preservation of Favoured Races in the Struggle for Life*. London: John Murray. [1, 6, 9, 12, 13, 14, 15, 20, 25, 29, 36, 37, 40, 43, 184, 187, 190]

Darwin, C. (1871). *The Descent of Man, and Selection in Relation to Sex*. 2 Volumes. London: John Murray. [192]

Darwin, C. (1872). *The Origin of Species by Means of Natural Selection, or the Preservation of Favoured Races in the Struggle for Life* (Sixth edn). London: John Murray. [14, 41, 43]

Davis, M., Hut, P. & Muller, R.A. (1984). Extinction of species by periodic comet showers. *Nature* **308**, 715–717. [30]

Davis, M.B. (1976). Pleistocene biogeography of temperate deciduous forests. *Geoscience and Man* **13**, 13–26. [42, 111, 112, 114]

Davis, M.B. (1981a). Outbreaks of forest pathogens in Quaternary history. In D.C. Bharadwaj, Vishnu-Mittre & H.K. Maheshwari (Eds.), *Fourth International Palynological Conference Proceedings Volume III*, pp. 216–228. Lucknow: Birbal Sahni Institute of Palaeobotany. [42]

Davis, M.B. (1981b). Quaternary history and the stability of forest communities. In D.C. West, H.H. Shugart & D.B. Botkin (Eds.), *Forest Succession: Concepts and Application*, pp. 132–153. New York: Springer. [111, 112, 114]

Davis, M.B. (1984). Holocene vegetational history of the eastern United States. In H.E. Wright, Jr (Ed.), *Late-Quaternary Environments of the United States. Vol 2. The Holocene*, pp. 166–181. London: Longman. [111, 112]

Davis, M.B., Woods, K.D., Webb, S.L. & Futyma, R.P. (1986). Dispersal versus climate: expansion of *Fagus* and *Tsuga* into the Upper Great Lakes region. *Vegetatio* **67**, 93–103. [112]

Dawkins, R. (1976). *The Selfish Gene*. Oxford: Oxford University Press. [26, 34]

Dawkins, R. (1983). *The Extended Phenotype: the Long Reach of the Gene*. Oxford: Oxford University Press. [26, 34]

Dawkins, R. (1986). *The Blind Watchmaker*. Harlow, England: Longman Scientific & Technical. [26, 34]

Dawkins, R. (1995). *River out of Eden: a Darwinian View of Life*. London: Weidenfeld & Nicholson. [26, 34]

de Beaulieu, J.L. & Reille, M. (1984). A long Upper Pleistocene pollen record from Les Echets, near Lyons, France. *Boreas* **13**, 111–132. [102]

de Beaulieu, J.L. & Reille, M. (1992). The last climatic cycle at La Grande Pile (Vosges, France). A new pollen profile. *Quaternary Science Reviews* **11**, 431–438. [102]

de Vries, H. (1906). *Species and Varieties: their Origin by Mutation* (Second edn). Chicago: Open Court. [14, 15, 29, 43]

Dean, W.E., Gardner, J.V. & Cepek, P. (1981). Tertiary carbonate-dissolution cycles on the Sierra Leone rise, eastern equatorial Atlantic Ocean. *Marine Geology* **39**, 81–101. [72]

Delcourt, P.A. & Delcourt, H.R. (1987). *Long-term Forest Dynamics of the Temperate Zone*, Ecological Studies No. 63. New York: Springer-Verlag. [114]

Dennett, D.C. (1995). *Darwin's Dangerous Idea: Evolution and the Meanings of Life*. London: Allen Lane. [12, 26, 196]

Denton, G.H. & Hughes, T.J. (1981). *The Last Great Ice Sheets*. New York: Wiley. [93]

Desmond, A. & Moore, J. (1991). *Darwin*. London: Michael Joseph. [170]

Dewar, R.E. (1984). Extinctions in Madagascar: the loss of the subfossil fauna. In P.S. Martin & R.G. Klein (Eds.), *Quaternary Extinctions: a Prehistoric Revolution*, pp. 574–593. Tucson, AZ: University of Arizona Press. [180, 181]

Dexter, F., Banks, H.T. & Webb, T., III (1987). Modeling Holocene changes in the location and abundance of beech populations in eastern North America. *Review of Palaeobotany and Palynology* **50**, 273–292. [145]

Dial, K.P. & Czaplewski, N.J. (1990). Do woodrat middens accurately represent the animals' environments and diets? The Woodhouse Mesa study. In J.L. Betancourt, T.R. van Devender & P.S. Martin (Eds.), *Packrat Middens: the Last 40,000 Years of Biotic Change*, pp. 43–58. Tucson, AZ: University of Arizona Press. [116]

Dobzhansky, T. (1937). *Genetics and the Origin of Species*. New York: Columbia University Press. [14, 15, 16, 19, 22, 25]

Dodson, J.R., Enright, N.J. & McLean, R.F. (1988). A late Quaternary vegetation history for northern New Zealand. *Journal of Biogeography* **15**, 647–656. [135]

Dransfield, J., Flenley, J.R., King, S.M., Harkness, D.D. & Rapu, S. (1984). A recently extinct palm from Easter Island. *Nature* **312**, 750–752. [182]

Edwards, R.L. & Gallup, C.D. (1993). Dating of the Devils Hole calcite vein. *Science* **259**, 1626. [66]

Egler, F.E. (1947). Arid southeast Oahu vegetation, Hawaii. *Ecological Monographs* **17**, 383–435. [37]

Eldredge, N. (1985). *Unfinished Synthesis: Biological Hierarchies and Modern Evolutionary Thought*. New York: Oxford University Press. [16, 22, 196]

Eldredge, N. (1986). *Time Frames: the Rethinking of Darwinian Evolution and the Theory of Punctuated Equilibria*. London: Heinemann. [27, 196]

Eldredge, N. (1989). *Macroevolutionary Dynamics: Species, Niches, and Adaptive Peaks*. New York: McGraw-Hill. [27, 31, 196]

Eldredge, N. (1992). Punctuated equilibria, rates of change, and large-scale entities in evolutionary systems. In A. Somit & S.A. Peterson (Eds.), *The Dynamics of Evolution: the Punctuated Equilibrium Debate in the Natural and Social Sciences*, pp. 103–120. Ithaca, NY: Cornell University Press. [38]

Eldredge, N. (1995). *Reinventing Darwin: the Great Evolutionary Debate.* London: Weidenfield and Nicholson. [26, 27, 175]

Eldredge, N. & Gould, S.J. (1972). Punctuated equilibria: an alternative to phyletic gradualism. In T.J.M. Schopf (Ed.), *Models in Paleobiology*, pp. 82–115. San Francisco: Freeman, Cooper. [15, 26, 27, 28, 31, 34, 43, 167, 188, 190, 191]

Elias, S.A. (1994). *Quaternary Insects and their Environments.* Washington: Smithsonian Institution Press. [137]

Elton, C. (1927). *Animal Ecology.* London: Sidgwick & Jackson. [36, 37, 152, 188, 194]

Emiliani, C. (1955). Pleistocene temperatures. *Journal of Geology* **63**, 538–578. [67]

Endler, J.A. (1986). *Natural Selection in the Wild.* Princeton, NJ: Princeton University Press. [168, 171, 173, 186, 187]

Eyles, N. (1993). Earth's glacial record and its tectonic setting. *Earth-Science Reviews* **35**, 1–248. [89]

Faegri, K. & Iversen, J. (1989). *Textbook of Pollen Analysis* (Fourth edn). Chichester: Wiley. [95]

Feldman, M.W. (1989). Discussion: ecology and evolution. In J. Roughgarden, R.M. May & S.A. Levin (Eds.), *Perspectives in Ecological Theory*, pp. 135–139. Princeton, NJ: Princeton University Press. [1]

Finley, R.B., Jr (1990). Woodrat ecology and behavior and the interpretation of paleomiddens. In J.L. Betancourt, T.R. van Devender & P.S. Martin (Eds.), *Packrat Middens: the Last 40,000 Years of Biotic Change*, pp. 28–42. Tucson, AZ: University of Arizona Press. [116]

Fisher, R.A. (1930). *The Genetical Theory of Natural Selection.* Oxford: Clarendon Press. [15, 21, 192, 194]

Flenley, J.R. (1979). *The Equatorial Rain Forest: a Geological History.* London: Butterworths. [127, 150]

Flenley, J.R. & King, S.M. (1984). Late Quaternary pollen records from Easter Island. *Nature* **307**, 47–50. [182]

Flenley, J.R., King, S.M., Jackson, J., Chew, C., Teller, J.T. & Prentice, M.E. (1991). The late Quaternary vegetational and climatic history of Easter Island. *Journal of Quaternary Science* **6**, 85–115. [182]

Follieri, M., Magri, D. & Sadori, L. (1986). Late Pleistocene *Zelkova* extinction in central Italy. *New Phytologist* **103**, 269–273. [102]

Follieri, M., Magri, D. & Sadori, L. (1988). 250,000-year pollen record from Valle di Castiglione (Roma). *Pollen et Spores* **30**, 329–356. [96, 97, 102]

Fryer, G., Greenwood, P.H. & Peake, J.F. (1983). Punctuated equilibria, morphological stasis and the palaeontological documentation of speciation: a biological appraisal of a case history in an African lake. *Biological Journal of the Linnean Society* **20**, 195–205. [164]

Fryer, G., Greenwood, P.H. & Peake, J.F. (1985). The demonstration of speciation in fossil molluscs and living fishes. *Biological Journal of the Linnean Society* **26**, 325–336. [164]

Fuji, N. (1983). Palynological study of 200-meter core samples from Lake Biwa, central Japan. I: The palaeovegetational and palaeoclimatic changes during the last 600,000 years. *Transactions and Proceedings of the Palaeontological Society of Japan* **132**, 230–252. [137]

Fuji, N. (1986). Palynological study of 200-meter core samples from Lake Biwa, central Japan. II: The palaeovegetational and palaeoclimatic changes during the ca. 250,000–100,000 years. *Transactions and Proceedings of the Palaeontological Society of Japan* **144**, 490–515. [137]

Fuller, J.L. (1995). *Holocene Forest Dynamics in Southern Ontario, Canada.* Ph. D. thesis, University of Cambridge. [151]

Futuyma, D.J. (1987). On the role of species in anagenesis. *American Naturalist* **130**, 465–473. [32]

Geikie, J. (1874). *The Great Ice Age and its Relation to the Antiquity of Man.* London: W. Isbister. [40]

Geitzenauer, K.R. (1972). The Pleistocene calcareous nannoplankton of the subantarctic Pacific Ocean. *Deep-Sea Research* **19**, 45–60. [147]

Ghil, M. (1994). Cryothermodynamics: the chaotic dynamics of paleoclimate. *Physica D* **77**, 130–159. [66]

Ghiselin, M.T. (1974). A radical solution to the species problem. *Systematic Zoology* **23**, 536–544. [30]

Gilbert, G.K. (1895). Sedimentary measurement of Cretaceous time. *Journal of Geology* **3**, 121–127. [74]

Gingerich, P.D. (1984). Pleistocene extinctions in the context of origination–extinction equilibria in cenozoic mammals. In P.S. Martin & R.G. Klein (Eds.), *Quaternary Extinctions: a Prehistoric Revolution*, pp. 211–222. Tucson, AZ: University of Arizona Press. [178, 179, 180]

Glancy, T.J., Jr, Barron, E.J. & Arthur, M.A. (1986). An initial study of the sensitivity of modeled Cretaceous climate to cyclical insolation forcing. *Paleoceanography* **1**, 523–537. [62, 63, 76, 92]

Gleason, H.A. (1926). The individualistic concept of the plant association. *Bulletin of the Torrey Botanical Club* **53**, 7–26. [37, 148, 194]

Glen, W. (1994). What the impact / volcanism / mass-extinction debates are about. In W. Glen (Ed.), *The Mass-extinction Debates: How Science Works in a Crisis*, pp. 7–38. Stanford, CA: Stanford University Press. [30]

Godwin, H. (1934). Pollen analysis. An outline of the problems and potentialities of the method. Part II. General applications of pollen analysis. *New Phytologist* **33**, 325–358. [42]

Goldhammer, R.K., Dunn, P.A. & Hardie, L.A. (1987). High frequency glacio-eustatic sealevel oscillations with Milankovitch characteristics recorded in middle Triassic platform carbonates in northern Italy. *American Journal of Science* **287**, 853–892. [78]

Goodwin, P.W. & Anderson, E.J. (1985). Punctuated aggradational cycles: a general hypothesis of episodic stratigraphic accumulation. *Journal of Geology* **93**, 515–533. [89, 90]

Gould, J. (1837). Remarks on a group of ground finches from Mr. Darwin's collection, with characters of the new species. *Proceedings of the Zoological Society of London* **5**, 4–7. [170]

Gould, S.J. (1965). Is uniformitarianism necessary? *American Journal of Science* **263**, 223–228. [26, 38]

References

Gould, S.J. (1969). An evolutionary microcosm: Pleistocene and Recent history of the land snail *P. (Poecilozonites)* in Bermuda. *Bulletin of the Museum of Comparative Zoology* **138**, 407–532. [161]

Gould, S.J. (1970). Coincidence of climatic and faunal fluctuations in Pleistocene Bermuda. *Science* **168**, 572–573. [140, 162, 173, 191]

Gould, S.J. (1982). Darwinism and the expansion of evolutionary theory. *Science* **216**, 380–387. [27]

Gould, S.J. (1985). The paradox of the first tier: an agenda for paleobiology. *Paleobiology* **11**, 2–12. [2, 29, 30, 189, 190]

Gould, S.J. (1987). *Time's Arrow, Time's Cycle: Myth and Metaphor in the Discovery of Geological Time*. Cambridge, MA: Harvard University Press. [9]

Gould, S.J. (1988). Prolonged stability in local populations of *Cerion agassizi* (Pleistocene–Recent) on Great Bahama Bank. *Paleobiology* **14**, 1–18. [162]

Gould, S.J. (1991). Abolish the Recent. *Natural History* **91**(5), 16–21. [5]

Gould, S.J. (1992). Punctuated equilibrium in fact and theory. In A. Somit & S.A. Peterson (Eds.), *The Dynamics of Evolution: the Punctuated Equilibrium Debate in the Natural and Social Sciences*, pp. 54–84. Ithaca, NY: Cornell University Press. [175]

Gould, S.J. (1993a). Modified grandeur. *Natural History* **102**(3), 14–20. [13]

Gould, S.J. (1993b). The inexorable logic of the punctuational paradigm: Hugo de Vries on species selection. In D.R. Lees & D. Edwards (Eds.), *Evolutionary Patterns and Processes*, Linnean Society Symposium Series No. 14, pp. 3–18. London: Academic Press. [14]

Gould, S.J. (1994). Ernst Mayr and the centrality of species. *Evolution* **48**, 31–35. [27]

Gould, S.J. & Eldredge, N. (1977). Punctuated equilibria: the tempo and mode of evolution reconsidered. *Paleobiology* **3**, 115–151. [27]

Gould, S.J. & Eldredge, N. (1993). Punctuated equilibrium comes of age. *Nature* **366**, 223–227. [175, 188]

Gould, S.J. & Vrba, E.S. (1982). Exaptation — a missing term in the science of form. *Paleobiology* **8**, 4–15. [190]

Gould, S.J. & Woodruff, D.S. (1986). Evolution and systematics of *Cerion* (Mollusca: Pulmonata) on New Providence Island: a radical revision. *Bulletin of the American Museum of Natural History* **182**, 389–490. [162]

Gould, S.J. & Woodruff, D.S. (1990). History as a cause of area effects: an illustration from *Cerion* on Great Inagua, Bahamas. *Biological Journal of the Linnean Society* **40**, 67–98. [163]

Graham, R.W. (1986). Response of mammalian communities to environmental changes during the late Quaternary. In J. Diamond & T.J. Case (Eds.), *Community Ecology*, pp. 300–313. New York: Harper & Row. [141, 142, 143, 144]

Graham, R.W. & Grimm, E.C. (1990). Effects of global climate change on the patterns of terrestrial biological communities. *Trends in Ecology and Evolution* **5**, 289–292. [142]

Graham, R.W. & Lundelius, E.L., Jr (1995). FAUNMAP: an electronic database documenting late Quaternary distribution of mammal species. Available from Illinois State Museum online at URL http://www.museum.state.il.us/research/faunmap/. [141]

Grant, B.R. & Grant, P.R. (1989). *Evolutionary Dynamics of a Natural Population: the Large Cactus Finch of the Galápagos.* Chicago: University of Chicago Press. [171, 173]

Grant, P.R. (1986). *Ecology and Evolution of Darwin's Finches.* Princeton, NJ: Princeton University Press. [170, 171, 172, 173]

Grant, V. (1985). *The Evolutionary Process: a Critical View of Evolutionary Theory.* New York: Columbia University Press. [34]

Grayson, D.K. (1984a). Nineteenth century explanations of Pleistocene extinctions: a review and analysis. In P.S. Martin & R.G. Klein (Eds.), *Quaternary Extinctions: a Prehistoric Revolution,* pp. 5–39. Tucson, AZ: University of Arizona Press. [178]

Grayson, D.K. (1984b). Explaining Pleistocene extinctions: thoughts on the structure of a debate. In P.S. Martin & R.G. Klein (Eds.), *Quaternary Extinctions: a Prehistoric Revolution,* pp. 807–823. Tucson, AZ: University of Arizona Press. [181]

Grotzinger, J.P. (1986a). Upward shallowing platform cycles: a response to 2.2 billion years of low-amplitude, high-frequency (Milankovitch band) sea-level oscillations. *Paleoceanography* **1**, 403–416. [84, 85, 87]

Grotzinger, J.P. (1986b). Cyclicity and paleoenvironmental dynamics, Rocknest platform, northwest Canada. *Geological Society of America Bulletin* **97**, 1208–1231. [84, 85]

Haffer, J. (1990). Geoscientific aspects of allopatric speciation. In G. Peters & R. Hutterer (Eds.), *Vertebrates in the Tropics,* pp. 45–60. Bonn: Museum Alexander Koenig. [33]

Haldane, J.B.S. (1932). *The Causes of Evolution.* London: Longmans, Green. [15, 22]

Hambrey, M.J. & Harland, W.B. (Eds.) (1981). *Earth's pre-Pleistocene Glacial Record.* Cambridge: Cambridge University Press. [88, 92, 93]

Hardie, L.A., Bosellini, A. & Goldhammer, R.K. (1986). Repeated subaerial exposure of subtidal carbonate platforms, Triassic, northern Italy: evidence for high frequency sea level oscillations on a 10^4 year scale. *Paleoceanography* **1**, 447–457. [78, 79]

Harland, W.B., Armstrong, R.L., Cox, A.V., Craig, L.E., Smith, A.G. & Smith, D.G. (1990). *A Geologic Time Scale 1989.* Cambridge: Cambridge University Press. [3, 4, 41, 71, 75, 78, 83]

Harper, J.L. (1967). A Darwinian approach to plant ecology. *Journal of Ecology* **55**, 247–270. [38, 39]

Hart, M.B. (1987). Orbitally induced cycles in the Chalk facies of the United Kingdom. *Cretaceous Research* **8**, 335–348. [76, 78]

Haynes, C.V., Jr, Eyles, C.H., Pavlish, L.A., Ritchie, J.C. & Rybak, M. (1989). Holocene palaeoecology of the eastern Sahara; Selima Oasis. *Quaternary Science Reviews* **8**, 109–136. [128, 129]

Hays, J.D., Imbrie, J. & Shackleton, N.J. (1976). Variations in the earth's orbit: pacemaker of the ice ages. *Science* **194**, 1121–1132. [31, 65, 66, 67, 68, 69, 70]

Hays, J.D., Lozano, J.D., Shackleton, N. & Irving, H. (1976). Reconstruction of the Atlantic and western Indian Ocean sectors of the 18,000 B.P. Antarctic Ocean. *Memoirs of the Geological Society of America* **145**, 337–372. [67]

Heckel, P.H. (1986). Sea-level curve for Pennsylvanian eustatic marine transgressive-regressive depositional cycles along midcontinent outcrop belt, North America. *Geology* **14**, 330–334. [82, 83]

Heisler, J. & Tremaine, S. (1989). How dating uncertainties affect the detection of periodicity in extinctions and craters. *Icarus* **77**, 213–219. [30]

Herbert, T.D. & D'Hondt, S.L. (1990). Precessional climate cyclicity in late Cretaceous – early Tertiary marine sediments: a high resolution chronometer of Cretaceous – Tertiary boundary events. *Earth and Planetary Science Letters* **99**, 263–275. [73, 74, 90]

Herbert, T.D. & Fischer, A.G. (1986). Milankovitch climatic origin of mid-Cretaceous black shale rhythms in central Italy. *Nature* **321**, 739–743. [74, 75, 76, 77]

Holman, J.A. (1976). Paleoclimatic implications of "ecologically incompatible", herpetological species (late Pleistocene: southeastern United States). *Herpetologica* **32**, 290–295. [142]

Honacki, J.H., Kinman, K.E. & Koeppl, J.W. (Eds.) (1982). *Mammal Species of the World: a Taxonomic and Geographic Reference*. Lawrence, KA: Allen Press and Association of Systematics Collections. [5]

Hooghiemstra, H. (1984). Vegetational and climatic history of the High Plains of Bogotá, Colombia: a continuous record of the last 3.5 million years. *Dissertationes Botanicae* **79**, 368pp. [125, 126, 127]

Hooghiemstra, H. (1989). Quaternary and upper-Pliocene glaciations and forest development in the tropical Andes: evidence from a long high-resolution pollen record from the sedimentary basin of Bogotá, Colombia. *Palaeogeography, Palaeoclimatology, Palaeoecology* **72**, 11–26. [125]

Hooghiemstra, H. & Sarmiento, G. (1991). Long continental pollen record from a tropical intermontane basin: Late Pliocene and Pleistocene history from a 540-metre core. *Episodes* **14**, 107–115. [125]

Hope, G. & Tulip, J. (1994). A long vegetation history from lowland Irian Jaya, Indonesia. *Palaeogeography, Palaeoclimatology, Palaeoecology* **109**, 385–398. [135, 136]

House, M.R. (1985). A new approach to an absolute timescale from measurements of orbital cycles and sedimentary microrhythms. *Nature* **315**, 721–725. [90]

Hubbard, A.E. & Gilinsky, N.L. (1992). Mass extinctions as statistical phenomena: an examination of the evidence using χ^2 tests and bootstrapping. *Paleobiology* **18**, 148–160. [30]

Hull, D.L. (1976). Are species really individuals? *Systematic Zoology* **25**, 174–191. [30]

Hull, D.L. (1978). A matter of individuality. *Philosophy of Science* **45**, 335–360. [30]

Hull, D.L. (1980). Individuality and selection. *Annual Review of Ecology and Systematics* **11**, 311–332. [30, 31, 152]

Huntley, B. (1988). Europe. In B. Huntley & T. Webb, III (Eds.), *Vegetation History*, Handbook of Vegetation Science No. 7, pp. 341–383. Dordrecht: Kluwer Academic. [103, 106]

Huntley, B. (1990a). Dissimilarity mapping between fossil and contemporary pollen spectra in Europe for the past 13,000 years. *Quaternary Research* **33**, 360–376. [103, 108]

Huntley, B. (1990b). European vegetation history: palaeovegetation maps from pollen data – 13000 yr BP to present. *Journal of Quaternary Science* **5**, 103–122. [103]

Huntley, B. & Birks, H.J.B. (1983). *An Atlas of Past and Present Pollen Maps for Europe 0–13,000 Years Ago*. Cambridge: Cambridge University Press. [42, 100, 102, 103, 104, 105]

Huntley, B. & Prentice, I.C. (1993). Holocene vegetation and climates of Europe. In H.E. Wright, Jr, J.E. Kutzbach, T. Webb, III, W.F. Ruddiman, F.A. Street-Perrott & P.J. Bartlein (Eds.), *Global Climates Since the Last Glacial Maximum*, pp. 136–168. Minneapolis, MN: University of Minnesota Press. [102]

Hutchinson, G.E. (1959). Homage to Santa Rosalia or why are there so many kinds of animals? *American Naturalist* **93**, 145–159. [38]

Huxley, J.S. (1942). *Evolution: the Modern Synthesis*. London: George Allen & Unwin. [12, 15, 20, 22, 25, 187]

Imbrie, J., Berger, A., Boyle, E.A., Clemens, S.C., Duffy, A., Howard, W.R., Kukla, G., Kutzbach, J., Martinson, D.G., McIntyre, A., Mix, A.C., Molfino, B., Morley, J.J., Peterson, L.C., Pisias, N.G., Prell, W.L., Raymo, M.E., Shackleton, N.J. & Toggweiler, J.R. (1993). On the structure and origin of major glaciation cycles 2. The 100,000–year cycle. *Paleoceanography* **8**, 699–735. [70]

Imbrie, J. & Imbrie, K.P. (1979). *Ice Ages: Solving the Mystery*. London: Macmillan. [44, 45, 48, 49, 65, 70, 85]

Imbrie, J. & Kipp, N.G. (1971). A new micropaleontological method for quantitative paleoclimatology: application to a late Pleistocene Caribbean core. In K.K. Turekian (Ed.), *The Late Cenozoic Glacial Ages*, pp. 71–181. New Haven, CT: Yale University Press. [147]

Imbrie, J., Mix, A.C. & Martinson, D.G. (1993). Milankovitch theory viewed from Devils Hole. *Nature* **363**, 531–533. [66]

Iversen, J. (1941). Landnam i Danmarks Stenalder. *Danmarks Geologiske Undersøgelse*. II Række **66**, 68pp. [42]

Iversen, J. (1960). Problems of the early post-glacial forest development in Denmark. *Danmarks Geologiske Undersøgelse*. IV Række **4**(3), 126pp. [103]

Jablonski, D. (1986). Background and mass extinctions: the alternation of macro-evolutionary regimes. *Science* **231**, 129–133. [190]

Jablonski, D. & Bottjer, D.J. (1991). Environmental patterns in the origins of higher taxa: the post-Paleozoic fossil record. *Science* **252**, 1831–1833. [174]

Jablonski, D., Flessa, K.W. & Valentine, J.W. (1985). Biogeography and paleobiology. *Paleobiology* **11**, 75–90. [152]

Jacobson, G.L., Jr, Webb, T., III & Grimm, E.C. (1987). Patterns and rates of vegetation change during the deglaciation of eastern North America. In W.F. Ruddiman & H.E. Wright, Jr (Eds.), *North America and Adjacent Oceans During the Last Deglaciation*, The Geology of North America Volume K-3, pp. 277–288. Boulder, CO: Geological Society of America. [113]

Jennings, L.J. (Ed.) (1884). *The Correspondence and Diaries of the Late Right Honourable John Wilson Croker, LL.D., F.R.S., Secretary to the Admiralty from 1809–1830*. 3 volumes. London: John Murray. []

Kareiva, P.M., Kingsolver, J.G. & Huey, R.B. (Eds.) (1993). *Biotic Interactions and Global Change*. Sunderland, MA: Sinauer Associates. [185]

References

Keltner, J. (1995). ShowTime. Software available from NOAA Paleoclimatology Program online at URL http://www.ngdc.noaa.gov/paleo/softlib.html. [111]

Kerney, M.P. & Cameron, R.A.D. (1979). *A Field Guide to the Land Snails of Britain and North-west Europe*. London: Collins. [141]

Kerney, M.P., Preece, R.C. & Turner, C. (1980). Molluscan and plant biostratigraphy of some late Devensian and Flandrian deposits in Kent. *Philosophical Transactions of the Royal Society of London Series B* **291**, 1–43. [140, 141]

Kershaw, A.P. (1978). Record of last interglacial-glacial cycle from northeastern Queensland. *Nature* **272**, 159–161. [132]

Kershaw, A.P. (1984). Late Cenozoic plant extinctions in Australia. In P.S. Martin & R.G. Klein (Eds.), *Quaternary Extinctions: a Prehistoric Revolution*, pp. 691–707. Tucson, AZ: University of Arizona Press. [181]

Kershaw, A.P. (1985). An extended late Quaternary vegetation record from northeastern Queensland and its implications for the seasonal tropics of Australia. *Proceedings of the Ecological Society of Australia* **13**, 179–189. [132]

Kershaw, A.P. (1986). Climatic change and Aboriginal burning in north-east Australia during the last two glacial/interglacial cycles. *Nature* **322**, 47–49. [131, 132]

Kurtén, B. & Anderson, E. (1980). *Pleistocene Mammals of North America*. New York: Columbia University Press. [157]

Kutzbach, J.E. (1981). Monsoon climate of the early Holocene: climate experiment with the Earth's orbital parameters for 9000 years ago. *Science* **214**, 59–61. [56]

Kutzbach, J.E. & Gallimore, R.G. (1988). Sensitivity of a coupled atmosphere/mixed ocean model to changes in orbital forcing at 9000 years B.P. *Journal of Geophysical Research* **93D**, 803–821. [51]

Kutzbach, J.E. & Gallimore, R.G. (1989). Pangaean climates: megamonsoons of the megacontinent. *Journal of Geophysical Research* **94D**, 3341–3357. [56, 61]

Kutzbach, J.E. & Guetter, P.J. (1986). The influence of changing orbital parameters and surface boundary conditions on climate simulations for the past 18 000 years. *Journal of the Atmospheric Sciences* **43**, 1726–1759. [51, 115, 133]

Kutzbach, J.E., Guetter, P.J., Behling, P.J. & Selin, R. (1993). Simulated climatic changes: results of the COHMAP climate-model experiments. In H.E. Wright, Jr, J.E. Kutzbach, T. Webb, III, W.F. Ruddiman, F.A. Street-Perrott & P.J. Bartlein (Eds.), *Global Climates Since the Last Glacial Maximum*, pp. 24–93. Minneapolis, MN: University of Minnesota Press. [51, 52, 57, 95, 130, 133]

Kutzbach, J.E., Guetter, P.J., Ruddiman, W.F. & Prell, W.L. (1989). Sensitivity of climate to late Cenozoic uplift in southern Asia and the American West: numerical experiments. *Journal of Geophysical Research* **94D**, 18393–18407. [87]

Kutzbach, J.E. & Otto-Bliesner, B.L. (1982). The sensitivity of the African-Asian monsoonal climate to orbital parameter changes for 9000 years B.P. in a low resolution general circulation model. *Journal of the Atmospheric Sciences* **39**, 1177–1188. [56, 130]

Labeyrie, L.D., Duplessy, J.C. & Blanc, P.L. (1987). Variations in mode of formation and temperature of oceanic deep waters over the past 125,000 years. *Nature* **327**, 477–482. [67]

Lack, D. (1947). *Darwin's Finches*. Cambridge: University Press. [170, 171]

Laskar, J. (1989). A numerical experiment on the chaotic behaviour of the Solar System. *Nature* **338**, 237–238. [46, 47]

Laskar, J. (1990). The chaotic motion of the Solar System: a numerical estimate of the size of the chaotic zones. *Icarus* **88**, 266–291. [46]

Lazarus, D. (1986). Tempo and mode of morphologic evolution near the origin of the radiolarian lineage *Pterocanium prismatium*. *Paleobiology* **12**, 175–189. [166]

Leopold, E.B. (1967). Late-Cenozoic patterns of plant extinction. In P.S. Martin & H.E. Wright, Jr (Eds.), *Pleistocene Extinctions: the Search for a Cause*, pp. 203–246. New Haven, CT: Yale University Press. [181]

Levinton, J. (1988). *Genetics, Paleontology, and Macroevolution*. Cambridge: Cambridge University Press. [34]

Leyden, B.W., Brenner, M., Hodell, D.A. & Curtis, J.H. (1994). Orbital and internal forcing of climate on the Yucatan Peninsula for the past ca. 36 ka. *Palaeogeography, Palaeoclimatology, Palaeoecology* **109**, 193–210. [123, 124]

Lister, A.M. (1984a). Evolutionary and ecological origins of British deer. *Proceedings of the the Royal Society of Edinburgh* **82B**, 205–229. [145, 156]

Lister, A.M. (1984b). The fossil record of elk (*Alces alces* (L.)) in Britain. *Quaternary Newsletter* **44**, 1–7. [157]

Lister, A.M. (1986). New results on deer from Swanscombe, and the stratigraphical significance of deer in the Middle and Upper Pleistocene of Europe. *Journal of Archaeological Science* **13**, 319–338. [145, 156]

Lister, A.M. (1989). Rapid dwarfing of red deer on Jersey in the last interglacial. *Nature* **342**, 539–542. [155, 156]

Lozhkin, A.V., Anderson, P.M., Eisner, W.R., Ravako, L.G., Hopkins, D.M., Brubaker, L.B., Colinvaux, P.A. & Miller, M.C. (1993). Late Quaternary lacustrine pollen records from southwestern Beringia. *Quaternary Research* **39**, 314–324. [137]

Ludwig, K.R., Simmons, K.R., Szabo, B.J., Winograd, I.J., Landwehr, J.M., Riggs, A.C. & Hoffman, R.J. (1992). Mass-spectrometric ^{230}Th–^{234}U–^{238}U dating of the Devils Hole calcite vein. *Science* **258**, 284–287. [65]

Ludwig, K.R., Simmons, K.R., Winograd, I.J., Szabo, B.J., Landwehr, J.M. & Riggs, A.C. (1993). Last interglacial in Devils Hole: reply. *Nature* **362**, 596. [66]

Lundelius, E.L., Jr (1983). Climatic implications of Late Pleistocene and Holocene faunal associations in Australia. *Alcheringa* **7**, 125–149. [142]

Lundelius, E.L., Jr, Graham, R.W., Anderson, E., Guilday, J., Holman, J.A., Steadman, D.W. & Webb, S.D. (1983). Terrestrial vertebrate faunas. In S.C. Porter (Ed.), *Late-Quaternary Environments of the United States. Vol 1. The Late Pleistocene*, pp. 311–353. London: Longman. [141, 142]

Lyell, C. (1826). On a recent formation of freshwater limestone in Forfarshire, and on some recent deposits of freshwater marl; with a comparison of recent with ancient freshwater formations; and an Appendix on the gyrogonite or seed-vessel of the Chara. *Transactions of the Geological Society Second Series* **II**, 73–96. [39]

Lyell, C. (1830). *Principles of Geology, being an Attempt to Explain the Former Changes of the Earth's Surface, by Reference to Causes Now in Operation*. Vol. I. London: John Murray. [7, 9]

Lyell, C. (1832). *Principles of Geology, being an Attempt to Explain the Former Changes of the Earth's Surface, by Reference to Causes Now in Operation.* Volume the second. London: John Murray. [7, 8, 9, 31, 187]

Lyell, C. (1833). *Principles of Geology, being an Attempt to Explain the Former Changes of the Earth's Surface, by Reference to Causes Now in Operation.* Vol. III. London: John Murray. [7, 9]

Lyell, C. (1875). *Principles of Geology or the Modern Changes of the Earth and its Inhabitants Considered as Illustrative of Geology* (Twelfth edn). 2 Volumes. London: John Murray. [10]

Mabberley, D.J. (1987). *The Plant-book: a Portable Dictionary of the Higher Plants.* Cambridge: Cambridge University Press. [5]

MacArthur, R.H. (1965). Patterns of species diversity. *Biological Reviews* **40**, 510–533. [39]

MacArthur, R.H. (1972). *Geographical Ecology: Patterns in the Distribution of Species.* New York: Harper & Row. [38, 39]

MacDonald, G.M. (1993). Fossil pollen analysis and the reconstruction of plant invasions. *Advances in Ecological Research* **24**, 67–109. [150]

MacDonald, G.M. & Cwynar, L.C. (1991). Post-glacial population growth rates of *Pinus contorta* ssp. *latifolia* in western Canada. *Journal of Ecology* **79**, 417–429. [151]

Maglio, V.J. (1978). Patterns of faunal evolution. In V.J. Maglio & H.B.S. Cooke (Eds.), *Evolution of African Mammals*, pp. 603–619. Cambridge, MA: Harvard University Press. [179]

Magri, D. (1989). Interpreting long-term exponential growth of plant populations in a 250000-year pollen record from Valle di Castiglione (Roma). *New Phytologist* **112**, 123–128. [96, 152]

Martin, P.S. (1984). Prehistoric overkill: the global model. In P.S. Martin & R.G. Klein (Eds.), *Quaternary Extinctions: a Prehistoric Revolution*, pp. 354–403. Tucson, AZ: University of Arizona Press. [181]

Martinson, D.G., Pisias, N.G., Hays, J.D., Imbrie, J., Moore, T.C., Jr & Shackleton, N.J. (1987). Age dating and the orbital theory of the ice ages: development of a high-resolution 0 to 300,000-year chronostratigraphy. *Quaternary Research* **27**, 1–29. [71, 178]

Matthews, J.V., Jr (1974). Quaternary environments at Cape Deceit (Seward Peninsula, Alaska): evolution of a tundra ecosystem. *Geological Society of America Bulletin* **85**, 1353–1384. [159]

Maynard Smith, J. (1975). *The Theory of Evolution* (Third edn). Harmondsworth, Middx: Penguin Books. [26, 34]

Mayr, E. (1942). *Systematics and the Origin of Species from the Viewpoint of a Zoologist.* New York: Columbia University Press. [14, 15, 17, 18, 19, 20, 24, 25, 27, 37, 43, 165, 173, 188, 191]

Mayr, E. (1954). Change of genetic environment and evolution. In J. Huxley, A.C. Hardy & E.B. Ford (Eds.), *Evolution as a Process*, pp. 157–180. London: George Allen & Unwin. [19, 27, 164]

Mayr, E. (1963). *Animal Species and Evolution.* Cambridge, MA: Belknap Press of Harvard University Press. [19, 27, 28, 173]

Mayr, E. (1980). Prologue: some thoughts on the history of the evolutionary synthesis. In E. Mayr & W.B. Provine (Eds.), *The Evolutionary Synthesis: Perspectives on the Unification of Biology*, pp. 1–48. Cambridge, MA: Harvard University Press. [16]

Mayr, E. (1982). *The Growth of Biological Thought: Diversity, Evolution, and Inheritance.* Cambridge, MA: Belknap Press of Harvard University Press. [6, 20]

Mayr, E. (1988). *Toward a New Philosophy of Biology: Observations of an Evolutionist.* Cambridge, MA: Harvard University Press. [20]

Mayr, E. (1993). *One Long Argument: Charles Darwin and the Genesis of Modern Evolutionary Thought.* London: Penguin Books. [12, 20, 21, 187]

McCrea, W.H. (1975). Ice ages and the galaxy. *Nature* **255**, 607–609. [89]

McCune, A.R. (1996). Biogeographic and stratigraphic evidence for rapid speciation in semionotid fishes. *Paleobiology* **22**, 34–48. [157, 191]

McGlone, M.S. (1985). Plant bigeography and the late Cenozoic history of New Zealand. *New Zealand Journal of Botany* **23**, 723–749. [135]

McGlone, M.S. (1988). New Zealand. In B. Huntley & T. Webb, III (Eds.), *Vegetation History*, Handbook of Vegetation Science No. 7, pp. 557–599. Dordrecht: Kluwer Academic. [135]

McGlone, M.S. (1989). The Polynesian settlement of New Zealand in relation to environmental and biotic changes. *New Zealand Journal of Ecology* **12**(Supplement), 115–129. [135]

McIntosh, R.P. (1985). *The Background of Ecology: Concept and Theory.* Cambridge: Cambridge University Press. [36]

McIntyre, A. (1967). Coccoliths as paleoclimatic indicators of Pleistocene glaciation. *Science* **158**, 1314–1317. [146]

McIntyre, A., Ruddiman, W.F. & Jantzen, R. (1972). Southward penetrations of the North Atlantic polar front: faunal and floral evidence of large-scale surface water mass movements over the last 225,000 years. *Deep-Sea Research* **19**, 61–77. [147]

Mildenhall, D.C. (1980). New Zealand late Cretaceous and Cenozoic plant biogeography: a contribution. *Palaeogeography, Palaeoclimatology, Palaeoecology* **31**, 197–233. [181]

Miller, W., III (1993). Models of recurrent fossil assemblages. *Lethaia* **26**, 182–183. [195]

Mommersteeg, H.J.P.M., Loutre, M.F., Young, R., Wijmstra, T.A. & Hooghiemstra, H. (1995). Orbital forced frequencies in the 975 000 year pollen record from Tenaghi Philippon (Greece). *Climate Dynamics* **11**, 4–24. [98]

Moore, T.C., Jr, Pisias, N.G. & Dunn, D.A. (1982). Carbonate time series of the Quaternary and late Miocene sediments in the Pacific Ocean: a spectral comparison. *Marine Geology* **46**, 217–233. [72]

Morgan, A.V. & Morgan, A. (1980). Faunal assemblages and distributional shifts of Coleoptera during the late Pleistocene in Canada and the northern United States. *Canadian Entomologist* **112**, 1105–1128. [138]

Morley, J.J. & Hays, J.D. (1979). *Cycladophora davisiana*: a stratigraphic tool for Pleistocene North Atlantic and interhemispheric correlation. *Earth and Planetary Science Letters* **44**, 383–389. [147]

Morley, J.J. & Hays, J.D. (1983). Oceanographic conditions associated with high abundances of the radiolarian *Cycladophora davisiana*. *Earth and Planetary Science Letters* **66**, 63–72. [147]

216 · References

Morley, J.J., Hays, J.D. & Robertson, J.H. (1982). Stratigraphic framework for the late Pleistocene in the northwest Pacific Ocean. *Deep-Sea Research* **29A**, 1485–1499. [147]

Muller, J. (1981). Fossil pollen records of extant angiosperms. *Botanical Review* **47**, 1–142. [126]

Olsen, H. (1990). Astronomical forcing of meandering river behaviour: Milankovitch cycles in Devonian of East Greenland. *Palaeogeography, Palaeoclimatology, Palaeoecology* **79**, 99–115. [82]

Olsen, P.E. (1984). Periodicity of lake-level cycles in the late Triassic Lockatong Formation of the Newark Basin. In A.L. Berger, J. Imbrie, J. Hays, G. Kukla & B. Saltzman (Eds.), *Milankovitch and Climate, Part 1*, pp. 129–146. New York: D. Reidel. [78, 79, 80, 81]

Olsen, P.E. (1986). A 40-million-year lake record of early Mesozoic orbital climatic forcing. *Science* **234**, 842–848. [78, 81]

Olsen, P.E., McCune, A.R. & Thomson, K.S. (1982). Correlation of the early Mesozoic Newark Supergroup by vertebrates, principally fishes. *American Journal of Science* **282**, 1–44. [78, 157]

Olson, S.L. (1973). Evolution of the rails of the South Atlantic islands (Aves: Rallidae). *Smithsonian Contributions to Zoology* **152**, 53pp. [180]

Olson, S.L. & James, H.F. (1982). Fossil birds from the Hawaiian Islands: evidence for wholesale extinction by man before western contact. *Science* **217**, 633–635. [180]

Overpeck, J.T., Webb, R.S. & Webb, T., III (1992). Mapping eastern North American vegetation change of the past 18 ka: No-analogs and the future. *Geology* **20**, 1071–1074. [115]

Overpeck, J.T., Webb, R.S. & Webb, T., III (1995). Analog vegetation maps for eastern North America. Available online from NOAA Paleoclimatology Program at URL http://www.ngdc.noaa.gov/paleo/vegmap.html. [115]

Park, J. & Herbert, T.D. (1987). Hunting for paleoclimatic periodicities in a geologic time series with an uncertain timescale. *Journal of Geophysical Research* **92B**, 14027–14040. [76]

Park, J. & Oglesby, R.J. (1991). Milankovitch rhythms in the cretaceous: a gcm modelling study. *Palaeogeography, Palaeoclimatology, Palaeoecology* **90**, 329–355. [62, 64]

Patterson, C. & Smith, A.B. (1989). Periodicity in extinctions: the role of systematics. *Ecology* **70**, 802–811. [30]

Peacock, J.D. (1989). Marine molluscs and late Quaternary environmental studies with particular reference to the late-glacial period in northwest Europe: a review. *Quaternary Science Reviews* **8**, 179–192. [148]

Peckham, M. (1959). *The Origin of Species by Charles Darwin: a Variorum Text*. Philadelphia, PA: University of Pennsylvania Press. [6]

Peltier, W.R. (1995). Time dependent topography through the glacial cycle. Animation and graphics available from U.S. Geological Survey Global Change Research Group online at URL http://geochange.er.usgs.gov/pub/sea_level/Core/raw/peltier/peltier_ice.html. [94]

Pennington, W. (1986). Lags in adjustment of vegetation to climate caused by the pace of soil development: evidence from Britain. *Vegetatio* **67**, 105–118. [42]

Peters, R.H. (1991). *A Critique for Ecology*. Cambridge: Cambridge University Press. [36]

Peterson, G.M. (1993). Vegetational and climatic history of the western former Soviet Union. In H.E. Wright, Jr, J.E. Kutzbach, T. Webb, III, W.F. Ruddiman, F.A. Street-Perrott & P.J. Bartlein (Eds.), *Global Climates Since the Last Glacial Maximum*, pp. 169–193. Minneapolis, MN: University of Minnesota Press. [102, 103]

Pianka, E.R. (1988). *Evolutionary Ecology* (Fourth edn). New York: Harper & Row. [38]

Pimm, S.L., Moulton, M.P. & Justice, L.J. (1995). Bird extinctions in the central Pacific. In J.H. Lawton & R.M. May (Eds.), *Extinction Rates*, pp. 75–87. Oxford: Oxford University Press. [180, 181]

Pisias, N.G., Shackleton, N.J. & Hall, M.A. (1985). Stable isotope and calcium carbonate records from hydraulic piston cored Hole 574A: high resolution records from the middle Miocene. *Initial Reports of the Deep Sea Drilling Project* **85**, 735–748. [72]

Potts, D.C. (1983). Evolutionary disequilibrium among Indo-Pacific corals. *Bulletin of Marine Science* **33**, 619–632. [148, 149, 173]

Potts, D.C. (1984). Generation times and the Quaternary evolution of reef-building corals. *Paleobiology* **10**, 48–58. [148, 164, 173]

Potts, R. & Behrensmeyer, A.K. (1992). Late cenozoic terrestrial ecosystems. In A.K. Behrensmeyer, J.D. Damuth, W.A. DiMichele, R. Potts, H.D. Sues & S.L. Wing (Eds.), *Terrestrial Ecosytems Through Time: Evolutionary Palaeoecology of Terrestrial Plants and Animals*, pp. 419–541. Chicago: University of Chicago Press. [197]

Prentice, I.C. (1988). Palaeoecology and plant population dynamics. *Trends in Ecology and Evolution* **3**, 343–345. [150]

Prentice, I.C., Bartlein, P.J. & Webb, T., III (1991). Vegetation and climate change in eastern North America since the last glacial maximum. *Ecology* **72**, 2038–2056. [113]

Pröbstl, M. & Grüger, E. (1979). Spätriss, Riss/Würm und Frühwürm am Samerberg im Oberbayern — ein vegetationsgeschichtlicher Beitrag zur Gliederung des Jungpleistozäns. *Geologica Bavarica* **80**, 5–64. [102]

Quinn, J.F. & Signor, P.W. (1989). Death stars, ecology and mass extinctions. *Ecology* **70**, 824–834. [30]

Raffi, S. (1986). The significance of marine boreal molluscs in the early Pleistocene faunas of the Mediterranean area. *Palaeogeography, Palaeoclimatology, Palaeoecology* **52**, 267–289. [147]

Rampino, M.R. & Stothers, R.B. (1984). Terrestrial mass extinctions, cometary impacts and the Sun's motion perpendicular to the galactic plane. *Nature* **308**, 709–712. [30]

Raup, D.M. & Sepkoski, J.J., Jr (1984). Periodicity of extinctions in the geologic past. *Proceedings of the National Academy of Sciences of the USA* **81**, 801–805. [29, 72]

Raup, D.M. & Sepkoski, J.J., Jr (1986). Periodic extinction of families and genera. *Science* **231**, 833–836. [29, 72]

Raup, D.M. & Sepkoski, J.J., Jr (1988). Testing for periodicity of extinction. *Science* **241**, 94–96. [29, 72]

References

Reille, M. & de Beaulieu, J.L. (1988). La fin de l'Eemian et les interstades du Prewürm mis pour la première fois en évidence dans le Massif Central français par l'analyse pollinique. *Comptes Rendus de l'Academie des Sciences Série II* **306**, 1205–1210. [102]

Research on Cretaceous Cycles Group (1986). Rhythmic bedding in Upper Cretaceous pelagic carbonate sequences: varying sedimentary responses to climatic forcing. *Geology* **14**, 153–156. [74]

Riddle, B.R., Honeycutt, R.L. & Lee, P.L. (1993). Mitochondrial DNA phylogeny in northern grasshopper mice (*Onychomys leucogaster*) — the influence of Quaternary climatic oscillations on population dispersal and divergence. *Molecular Ecology* **2**, 183–193. [144]

Ridley, M. (1993). *Evolution*. Boston, MA: Blackwell Scientific Publications. [35]

Ridley, M. (1994). *The Red Queen: Sex and the Evolution of Human Nature*. London: Penguin Books. [26, 192]

Ritchie, J.C. (1987a). *Postglacial Vegetation of Canada*. Cambridge: Cambridge University Press. [115]

Ritchie, J.C. (1987b). A Holocene pollen record from Bir Atrun, northwest Sudan. *Pollen et Spores* **29**, 391–410. [128, 130]

Ritchie, J.C. (1995). Current trends in studies of long-term plant community dynamics. *New Phytologist* **130**, 469–494. [120]

Ritchie, J.C., Eyles, C.H. & Haynes, C.V. (1985). Sediment and pollen evidence for an early to mid-Holocene humid period in the eastern Sahara. *Nature* **314**, 352–355. [128]

Ritchie, J.C. & Haynes, C.V. (1987). Holocene vegetation zonation in the eastern Sahara. *Nature* **330**, 645–647. [128, 129, 130]

Ritchie, J.C. & MacDonald, G.M. (1986). The patterns of post-glacial spread of white spruce. *Journal of Biogeography* **13**, 527–540. [115]

Robinson, N.D. (1986). Fining-upward microrhythms with basal scours in the Chalk of Kent and Surrey, England and their stratigraphic importance. *Newsletters on Stratigraphy* **17**, 21–28. [78]

Ruddiman, W.F. (1990). Changes in climate and biota on geologic time scales. *Trends in Ecology and Evolution* **5**, 285–288. [89]

Ruddiman, W.F. & Kutzbach, J.E. (1989). Forcing of late Cenozoic northern hemisphere climate by plateau uplift in southern Asia and the American West. *Journal of Geophysical Research* **94D**, 18409–18427. [88, 89]

Ruddiman, W.F. & Kutzbach, J.E. (1990). Late Cenozoic plateau uplift and climate change. *Transactions of the Royal Society of Edinburgh: Earth Sciences* **81**, 301–314. [87]

Ruddiman, W.F., Prell, W.L. & Raymo, M.E. (1989). Late Cenozoic uplift in southern Asia and the American West: rationale for general circulation model experiments. *Journal of Geophysical Research* **94D**, 18379–18391. [87]

Ruddiman, W.F. & Raymo, M.E. (1988). Northern hemisphere climate regimes during the past 3 Ma: possible tectonic connections. *Philosophical Transactions of the Royal Society of London Series B* **318**, 411–430. [71, 87]

Ruddiman, W.F., Raymo, M. & McIntyre, A. (1986). Matuyama 41,000-year cycles: North Atlantic Ocean and northern hemishpere ice sheets. *Earth and Planetary Science Letters* **80**, 117–129. [71]

Rudwick, M.J.S. (Ed.) (1990). *Principles of Geology by Charles Lyell*. Vol. 1. Chicago: Chicago University Press. [8]
Sancetta, C. & Robinson, S.W. (1983). Diatom evidence on Wisconsin and Holocene events in the Bering Sea. *Quaternary Research* **20**, 232–245. [147]
Schwarzacher, W. (1993). *Cyclostratigraphy and the Milankovitch Theory*, Developments in Sedimentology No. 52. Amsterdam: Elsevier. [65, 66]
Semken, H.A., Jr (1984). Holocene mammalian biogeography and climatic change in the eastern and central United States. In H.E. Wright, Jr (Ed.), *Late-Quaternary Environments of the United States. Vol 2. The Holocene*, pp. 182–207. London: Longman. [141, 142]
Sepkoski, J.J., Jr (1989). Periodicity in extinctions and the problem of catastrophism in the history of life. *Journal of the Geological Society, London* **146**, 7–19. [29, 72]
Sergin, V.Y. (1979). Numerical modeling of the glaciers–ocean–atmosphere global system. *Journal of Geophysical Research* **84C**, 3191–3204. [185]
Sergin, V.Y. (1980). Origin and mechanism of large-scale climatic oscillations. *Science* **209**, 1477–1482. [185]
Shackleton, N.J. (1967). Oxygen isotope analyses and Pleistocene temperatures re-assessed. *Nature* **215**, 15–17. [67]
Shackleton, N.J. (1989). Rosetta stone for Quaternary ice ages. *Current Contents: Physical, Chemical & Earth Sciences* **29**(5), 15. [67]
Shackleton, N.J. (1993a). The climate system in the recent geological past. *Philosophical Transactions of the Royal Society of London Series B* **341**, 209–213. [64]
Shackleton, N.J. (1993b). Last interglacial in Devils Hole. *Nature* **362**, 596. [66]
Shackleton, N.J. & Imbrie, J. (1990). The $\delta^{18}O$ spectrum of oceanic deep water over a five-decade band. *Climatic Change* **16**, 217–230. [72, 73]
Shackleton, N.J., Imbrie, J. & Pisias, N.G. (1988). The evolution of oceanic oxygen-isotope variability in the North Atlantic over the past three million years. *Philosophical Transactions of the Royal Society of London Series B* **318**, 679–688. [71]
Shackleton, N.J. & Opdyke, N.D. (1973). Oxygen isotope and palaeomagnetic stratigraphy of equatorial Pacific core V28-238: oxygen isotope temperatures and ice volumes on a 10^5 year and 10^6 year scale. *Quaternary Research* **3**, 39–55. [5, 40, 67, 133, 178]
Shackleton, N.J. & Opdyke, N.D. (1976). Oxygen-isotope and paleomagnetic stratigraphy of Pacific core V28-239 late Pliocene to latest Pleistocene. *Geological Society of America Memoir* **145**, 449–464. [99]
Shaw, H.R. (1994). *Craters, Cosmos, and Chronicles: a New Theory of the Earth*. Stanford, CA: Stanford University Press. [89]
Shoemaker, E.M., Wolfe, R.F. & Shoemaker, C.S. (1990). Asteroid and comet flux in the neighbourhood of the Earth. In V.L. Sharpton & P.D. Ward (Eds.), *Global Catastrophes in Earth history: an Interdisciplinary Conference on Impacts, Volcanism, and Mass Mortality*, Geological Society of America Special Paper No. 247, pp. 155–170. [30]
Sibley, C.G. & Monroe, B.L., Jr (1990). *Distribution and Taxonomy of Birds of the World*. New Haven, CT: Yale University Press. [170]
Simpson, G.G. (1944). *Tempo and Mode in Evolution*, Columbia Biological Series No. XV. New York: Columbia University Press. [14, 15, 20, 21, 22, 23, 25, 27, 173, 189]

Singh, G. & Geissler, E.A. (1985). Late Cainozoic history of vegetation, fire, lake levels and climate, at Lake George, New South Wales, Australia. *Philosophical Transactions of the Royal Society of London Series B* **311**, 379–447. [133, 134]

Singh, G., Kershaw, A.P. & Clark, R. (1981). Quaternary vegetation and fire history in Australia. In A.M. Gill, R.A. Groves & I.R. Noble (Eds.), *Fire and the Australian Biota*, pp. 23–54. Canberra: Australian Academy of Science. [132]

Singh, G., Opdyke, N.D. & Bowler, J.M. (1981). Late Cainozoic stratigraphy, palaeomagnetic chronology and vegetational history from Lake George, N.S.W. *Journal of the Geological Society of Australia* **28**, 435–452. [133]

Smart, P.L. & Frances, P.D. (Eds.) (1991). *Quaternary Dating Methods — a User's Guide*, Technical Guide No. 4. Cambridge: Quaternary Research Association. [41]

Smit, A. & Wijmstra, T.A. (1970). Application of transmission electron microscope analysis to the reconstruction of former vegetation. *Acta Botanica Neerlandica* **19**, 867–876. [99]

Sondaar, P.Y. (1977). Insularity and its effect on mammal evolution. In M.K. Hecht, P.C. Goody & B.M. Hecht (Eds.), *Major Patterns in Vertebrate Evolution*, pp. 671–707. New York: Plenum Press. [155, 181]

Sorhannus, U., Fenster, E.J., Burckle, L.H. & Hoffman, A. (1988). Cladogenetic and anagenetic changes in the morphology of *Rhizosolenia praebergonii* Mukhina. *Historical Biology* **1**, 185–205. [166, 167]

Spaulding, W.G. (1990). Vegetational and climatic development of the Mojave Desert: the last glacial maximum to the present. In J.L. Betancourt, T.R. van Devender & P.S. Martin (Eds.), *Packrat Middens: the Last 40,000 Years of Biotic Change*, pp. 166–199. Tucson, AZ: University of Arizona Press. [117]

Spaulding, W.G., Betancourt, J.L., Croft, L.K. & Cole, K.L. (1990). Packrat middens: their composition and methods of analysis. In J.L. Betancourt, T.R. van Devender & P.S. Martin (Eds.), *Packrat Middens: the Last 40,000 Years of Biotic Change*, pp. 59–84. Tucson, AZ: University of Arizona Press. [116]

Spaulding, W.G., Leopold, E.B. & van Devender, T.R. (1983). Late Wisconsin paleoecology of the American southwest. In S.C. Porter (Ed.), *Late-Quaternary Environments of the United States. Vol 1. The Late Pleistocene*, pp. 259–293. London: Longman. [117, 118]

Stanley, S.M. (1975). A theory of evolution above the species level. *Proceedings of the National Academy of Sciences of the USA* **72**, 646–650. [27, 29, 31]

Stanley, S.M. (1979). *Macroevolution: Pattern and Process*. San Francisco: W. H. Freeman. [174, 179]

Stanley, S.M. (1985). Rates of evolution. *Paleobiology* **11**, 13–26. [174, 175, 179]

Stanley, S.M. (1989). Fossils, macroevolution, and theoretical ecology. In J. Roughgarden, R.M. May & S.A. Levin (Eds.), *Perspectives in Ecological Theory*, pp. 125–134. Princeton, NJ: Princeton University Press. [1]

Steadman, D.W. (1986). Holocene vertebrate fossils from Isla Floreana, Galápagos. *Smithsonian Contributions to Zoology* **413**, 103pp. [177, 180]

Steadman, D.W. (1989). Extinction of birds in eastern Polynesia: a review of the record, and comparison with other Pacific island groups. *Journal of Archaeological Science* **16**, 177–205. [180]

Steadman, D.W. & Martin, P.S. (1984). Extinction of birds in the late Pleistocene of North America. In P.S. Martin & R.G. Klein (Eds.), *Quaternary Extinctions: a Prehistoric Revolution*, pp. 466–477. Tucson, AZ: University of Arizona Press. [180]

Steadman, D.W., Pregill, G.K. & Olson, S.L. (1984). Fossil vertebrates from Antigua, Lesser Antilles: evidence for late Holocene human-caused extinctions in the West Indies. *Proceedings of the National Academy of Sciences of the USA* **81**, 4448–4451. [180]

Stebbins, G.L., Jr (1950). *Variation and Evolution in Plants*. New York: Columbia University Press. [14, 15, 22, 24, 192]

Stebbins, G.L. (1972). Ecological distribution of centers of major adaptive radiation in Angiosperms. In D.H. Valentine (Ed.), *Taxonomy Phytogeography and Evolution*, pp. 7–34. London: Academic Press. [26]

Stebbins, G.L. (1984). Polyploidy and the distribution of the arctic-alpine flora: new evidence and a new approach. *Botanica Helvetica* **94**, 1–13. [24, 191]

Stigler, S.M. & Wagner, M.J. (1987). A substantial bias in nonparametric tests for periodicity in geophysical data. *Science* **215**, 1501–1503. [30]

Stigler, S.M. & Wagner, M.J. (1988). Response. *Science* **241**, 96–99. [30]

Street-Perrott, F.A. (1991). General circulation (GCM) modelling of palaeoclimates: a critique. *The Holocene* **1**, 74–80. [51]

Streeter, S.S. & Shackleton, N.J. (1979). Palaeocirculation of the deep North Atlantic: 150,000-year record of benthic foraminifera and oxygen-18. *Science* **203**, 168–171. [147]

Stuart, A.J. (1982). *Pleistocene Vertebrates in the British Isles*. London: Longman. [154, 157]

Stuiver, M. & Reimer, P.J. (1993). Extended ^{14}C data base and revised CALIB 3.0 ^{14}C age calibration program. *Radiocarbon* **35**, 215–230. [4]

Sulloway, F.J. (1982a). Darwin and his finches: the evolution of a legend. *Journal of the History of Biology* **15**, 1–53. [170]

Sulloway, F.J. (1982b). darwin's conversion: the Beagle voyage and its aftermath. *Journal of the History of Biology* **15**, 325–396. [170]

Sulloway, F.J. (1982c). The Beagle collections of Darwin's finches (Geospizinae). *Bulletin of the British Museum (Natural History) Zoology Series* **43**, 49–94. [170]

Sussman, G.J. & Wisdom, J. (1992). Chaotic evolution of the Solar System. *Science* **257**, 56–62. [47]

Thompson, R.S. (1990). Late Quaternary vegetation and climate in the Great Basin. In J.L. Betancourt, T.R. van Devender & P.S. Martin (Eds.), *Packrat Middens: the Last 40,000 Years of Biotic Change*, pp. 200–239. Tucson, AZ: University of Arizona Press. [117, 119]

Thomsen, E. & Vorren, T.O. (1986). Macrofaunal palaeoecology and stratigraphy in late Quaternary shelf sediments off northern Norway. *Palaeogeography, Palaeoclimatology, Palaeoecology* **56**, 103–150. [148]

Thomson, W. (1862). On the secular cooling of the Earth. *Transactions of the Royal Society of Edinburgh* **23**, 157–169. [40, 41]

Tiwari, R.K. (1987). Higher-order eccentricity cycles of the middle and late Miocene climatic variations. *Nature* **327**, 219–221. [72]

Traverse, A. (1988). Plant evolution dances to a different beat: plant and animal evolutionary mechanisms compared. *Historical Biology* **1**, 277–301. [30]

222 · References

Travis, J. & Mueller, L.D. (1989). Blending ecology and genetics: progress toward a unified population biology. In J. Roughgarden, R.M. May & S.A. Levin (Eds.), *Perspectives in Ecological Theory*, pp. 101–124. Princeton, NJ: Princeton University Press. [1]

Trotter, M.M. & McCulloch, B. (1984). Moas, men, and middens. In P.S. Martin & R.G. Klein (Eds.), *Quaternary Extinctions: a Prehistoric Revolution*, pp. 708–727. Tucson, AZ: University of Arizona Press. [180]

Tsukada, M. (1982a). *Pseudotsuga menziesii* (Mirb.) Franco: its pollen dispersal and late Quaternary history in the Pacific northwest. *Japanese Journal of Ecology* **32**, 159–187. [151]

Tsukada, M. (1982b). Late-Quaternary development of the *Fagus* forest in the Japanese archipelago. *Japanese Journal of Ecology* **32**, 113–118. [151]

Tsukada, M. (1985). Map of vegetation during the last glacial maximum in Japan. *Quaternary Research* **23**, 369–381. [137]

Tsukada, M. (1988). Japan. In B. Huntley & T. Webb, III (Eds.), *Vegetation History*, Handbook of Vegetation Science No. 7, pp. 459–518. Dordrecht: Kluwer Academic Publishers. [137]

Tsukada, M. & Sugita, S. (1982). Late Quaternary dynamics of pollen influx at Mineral Lake, Washington. *Botanical Magazine, Tokyo* **95**, 401–418. [151]

Tzedakis, P.C. (1993). Long-term tree populations in northwest Greece through multiple Quaternary climatic cycles. *Nature* **364**, 437–440. [100, 101]

Tzedakis, P.C. (1994). Vegetation change through glacial-interglacial cycles: a long pollen perspective. *Philosophical Transactions of the Royal Society of London Series B* **345**, 403–432. [100, 101]

Tzedakis, P.C. & Bennett, K.D. (1995). Interglacial vegetation succession: a view from southern Europe. *Quaternary Science Reviews* **14**, 967–982. [101]

Valentine, J.W. & Jablonski, D. (1991). Biotic effects of sea level change: the Pleistocene test. *Journal of Geophysical Research* **96B**, 6873–6878. [148]

Valentine, J.W. & Jablonski, D. (1993). Fossil communities: compositional variation at many time scales. In R.E. Ricklefs & D. Schluter (Eds.), *Species Diversity in Ecological Communities: Historical and Geographical Perspectives*, pp. 341–349. Chicago: University of Chicago Press. [197]

van Campo, E., Cour, P. & Sixuan, H. (1996). Holocene environmental changes in Bangong Co basin (Western Tibet). Part 2: The pollen record. *Palaeogeography, Palaeoclimatology, Palaeoecology* **120**, 49–63. [137]

van Campo, E. & Gasse, F. (1993). Pollen- and diatom-inferred climatic and hydrological changes in Sumxi Co Basin (western Tibet) since 13,000 yr B.P. *Quaternary Research* **39**, 300–313. [137]

van der Kaars, W.A. (1991). Palynology of eastern Indonesian marine piston-cores: a late Quaternary vegetational and climatic record for Australasia. *Palaeogeography, Palaeoclimatology, Palaeoecology* **85**, 239–302. [135]

van der Kaars, W.A. & Dam, M.A.C. (1995). A 135,000-year record of vegetational and climatic change from the Bandung area, West-Java, Indonesia. *Palaeogeography, Palaeoclimatology, Palaeoecology* **117**, 55–72. [137]

van der Wiel, A.M. & Wijmstra, T.A. (1987a). Palynology of the lower part (78–120m) of the core Tenaghi Philippon II, Middle Pleistocene of Macedonia, Greece. *Review of Palaeobotany and Palynology* **52**, 73–88. [96, 98]

van der Wiel, A.M. & Wijmstra, T.A. (1987b). Palynology of the 112.8–197.8m interval of the core Tenaghi Philippon III, Middle Pleistocene of Macedonia, Greece. *Review of Palaeobotany and Palynology* **52**, 89–117. [96, 98]

van Devender, T.R. (1990a). Late Quaternary vegetation and climate of the Chihuahuan Desert, United States and Mexico. In J.L. Betancourt, T.R. van Devender & P.S. Martin (Eds.), *Packrat Middens: the Last 40,000 Years of Biotic Change*, pp. 104–134. Tucson, AZ: University of Arizona Press. [117]

van Devender, T.R. (1990b). Late Quaternary vegetation and climate of the Sonoran Desert, United States and Mexico. In J.L. Betancourt, T.R. van Devender & P.S. Martin (Eds.), *Packrat Middens: the Last 40,000 Years of Biotic Change*, pp. 135–165. Tucson, AZ: University of Arizona Press. [117]

van Devender, T.R., Toolin, L.J. & Burgess, T.L. (1990). The ecology and paleoecology of grasses in selected Sonoran Desert plant communities. In J.L. Betancourt, T.R. van Devender & P.S. Martin (Eds.), *Packrat Middens: the Last 40,000 Years of Biotic Change*, pp. 326–349. Tucson, AZ: University of Arizona Press. [121]

van Houten, F.B. (1962). Cyclic sedimentation and the origin of analcime-rich upper Triassic Lockatong Formation, west-central New Jersey and adjacent Pennsylvania. *American Journal of Science* **260**, 561–576. [78]

van Houten, F.B. (1986). Search for Milankovitch patterns among oolitic ironstones. *Paleoceanography* **1**, 459–466. [66]

Van Valen, L. (1973). A new evolutionary law. *Evolutionary Theory* **1**, 1–30. [26]

Vaught, K.C. (1989). *A Classification of the Living Mollusca*. Melbourne, FL: American Malacologists. [5]

Vernekar, A.D. (1972). Long-period global variations of incoming solar radiation. *Meteorological Monographs* **12**, 21pp. [45, 67]

von Post, L. (1946). The prospect for pollen analysis in the study of the earth's climatic history. *New Phytologist* **45**, 193–217. [42]

Vrba, E.S. (1985). Environment and evolution: alternative causes of the temporal distribution of evolutionary events. *South African Journal of Science* **81**, 229–236. [31, 32, 174, 194]

Vrba, E.S. (1992). Mammals as key to evolutionary theory. *Journal of Mammalogy* **73**, 1–28. [31]

Vrba, E.S. (1993). Turnover-pulses, the Red Queen, and related topics. *American Journal of Science* **293-A**, 418–452. [31, 32, 33, 174, 194]

Vrba, E.S. & Gould, S.J. (1986). The hierarchical expansion of sorting and selection: sorting and selection cannot be equated. *Paleobiology* **12**, 217–228. [176]

Wallace, A.R. (1859). On the tendency of varieties to depart indefinitely from the original type. *Journal of the Proceedings of the Linnean Society. Zoology* **3**, 53–62. [11]

Ward, W.R. (1979). Present obliquity oscillations of Mars: fourth order accuracy in orbital e and I. *Journal of Geophysical Research* **84B**, 237–241. [48]

Wardle, P. (1988). Effects of glacial climates on floristic distribution in New Zealand 1. A review of the evidence. *New Zealand Journal of Botany* **26**, 541–555. [135]

Watts, W.A. (1961). Post Atlantic forests in Ireland. *Proceedings of the Linnean Society of London* **172**, 33–38. [42]

Weaver, J.E. & Clements, F.E. (1929). *Plant Ecology.* New York: McGraw-Hill. [37]

Webb, R.H. & Betancourt, J.L. (1990). The spatial and temporal distribution of radiocarbon ages from packrat middens. In J.L. Betancourt, T.R. van Devender & P.S. Martin (Eds.), *Packrat Middens: the Last 40,000 Years of Biotic Change*, pp. 85–102. Tucson, AZ: University of Arizona Press. [116]

Webb, S.D. (1984). Ten million years of mammal extinctions in North America. In P.S. Martin & R.G. Klein (Eds.), *Quaternary Extinctions: a Prehistoric Revolution*, pp. 189–210. Tucson, AZ: University of Arizona Press. [178, 179]

Webb, T., III (1986). Is vegetation in equilibrium with climate? How to interpret late-Quaternary pollen data. *Vegetatio* **67**, 75–91. [113]

Webb, T., III (1987). The appearance and disappearance of major vegetational assemblages: long-term vegetational dynamics in eastern North America. *Vegetatio* **69**, 177–187. [112]

Webb, T., III (1988). Eastern north America. In B. Huntley & T. Webb, III (Eds.), *Vegetation History*, Handbook of Vegetation Science No. 7, pp. 385–414. Dordrecht: Kluwer Academic. [112]

Webb, T., III & Bryson, R.A. (1972). Late- and postglacial climatic change in the northern midwest, USA: quantitative estimates derived from fossil pollen spectra by multivariate statistical analysis. *Quaternary Research* **2**, 70–115. [42]

Weedon, G.P. (1986). Hemipelagic shelf sedimentation and climatic cycles: the basal Jurassic (Blue Lias) of south Britain. *Earth and Planetary Science Letters* **76**, 321–335. [78]

Wei, K.Y. & Kennett, J.P. (1988). Phyletic gradualism and punctuated equilibrium in the late Neogene planktonic foraminiferal clade *Globoconella*. *Paleobiology* **14**, 345–363. [165]

Weiner, J. (1994). *The Beak of the Finch: Evolution in Real Time.* London: Jonathan Cape. [171]

Wells, P.V. & Jorgensen, C.D. (1964). Pleistocene wood rat middens and climatic change in Mohave Desert: a record of juniper woodlands. *Science* **143**, 1171–1174. []

West, R.G. (1964). Inter-relations of ecology and Quaternary palaeobotany. *Journal of Ecology* **52**(Supplement), 47–57. [103, 148]

West, R.G. (1980). Pleistocene forest history in East Anglia. *New Phytologist* **85**, 571–622. [95]

White, F. (1983). *The Vegetation of Africa: a Descriptive Memoir to Accompany the Unesco / AETFAT / UNSO Vegetation Map of Africa.* Paris: Unesco. [128]

Wijmstra, T.A. (1969). Palynology of the first 30 metres of a 120m deep section in northern Greece. *Acta Botanica Neerlandica* **18**, 511–527. [96, 98]

Wijmstra, T.A. & Smit, A. (1976). Palynology of the middle part (30–78 metres) of the 120m deep section in northern Greece (Macedonia). *Acta Botanica Neerlandica* **25**, 297–312. [96, 98]

Williams, G.C. (1966). *Adaptation and Natural Selection: a Critique of Some Current Evolutionary Thought.* Princeton, NJ: Princeton University Press. [26]

Williams, G.C. (1992). *Natural Selection: Domains, Levels, and Challenges.* New York: Oxford University Press. [26, 184, 196]

Williams, G.E. (1975). Possible relation between periodic glaciation and the flexure of the galaxy. *Earth and Planetary Science Letters* **26**, 361–369. [89]

Williams, G.E. (1991). Milankovitch-band cyclicity in bedded halite deposits contemporaneous with Late Ordovician–Early Silurian glaciation, Canning Basin, Western Australia. *Earth and Planetary Science Letters* **103**, 143–155. [84, 87]

Williamson, P.G. (1981). Palaeontological documentation of speciation in Cenozoic molluscs from Turkana basin. *Nature* **293**, 437–443. [163, 164]

Williamson, P.G. (1985a). Punctutated equilibrium, morphological stasis and the palaeontological documentation of speciation: a reply to Fryer, Greenwood and Peake's critique of the Turkana Basin mollusc sequence. *Biological Journal of the Linnean Society* **26**, 307–324. [164]

Williamson, P.G. (1985b). In reply to Fryer, Greenwood and Peake. *Biological Journal of the Linnean Society* **26**, 337–340. [164]

Willis, K.J. (1994). The vegetational history of the Balkans. *Quaternary Science Reviews* **13**, 769–788. [102]

Willis, K.J. & Bennett, K.D. (1995). Mass extinction, punctuated equilibrium and the fossil plant record. *Trends in Ecology and Evolution* **10**, 308–309. [30]

Willson, M.F. & Burley, N. (1983). *Mate Choice in Plants: Tactics, Mechanisms, and Consequences*, Monographs in Population Biology No. 19. Princeton, NJ: Princeton University Press. [192]

Wilson, E.O. (1994). *The Diversity of Life*. London: Penguin Books. [150]

Wilson, L.G. (1970). *Sir Charles Lyell's Scientific Journals on the Species Question*. New Haven, CT: Yale University Press. [9]

Winograd, I.J., Coplen, T.B., Landwehr, J.M., Riggs, A.C., Ludwig, K.R., Szabo, B.J., Kolesar, P.T. & Revesz, K.M. (1992). Continuous 500,000-year climate record from vein calcite in Devils Hole, Nevada. *Science* **258**, 255–260. [65]

Woillard, G.M. (1978). Grande Pile peat bog: a continuous pollen record for the last 140,000 years. *Quaternary Research* **9**, 1–21. [102]

Worster, D. (1985). *Nature's Economy: a History of Ecological Ideas*. Cambridge: Cambridge University Press. [8]

Wright, S. (1931). Evolution in Mendelian populations. *Genetics* **16**, 97–159. [15, 21, 194]

Index

Page numbers are in normal type for text references, **bold** type for references in Tables, *italic* type for references in Figures, and underlined for references to quotations.

Abies, fir trees (Pinaceae), 100, 111, **151, 152**, *169*
   pollen, *75, 98, 101, 110*
Abies balsamea, balsam fir (North America)
   pollen, *112*
Abies concolor, white fir (North America), 120, 122
   macrofossils, *118*
Abies lasiocarpa, subalpine fir (North America), 122
Acacia, trees and shrubs (Leguminosae), 128, 129, 181
   pollen, *130*
Acer, sycamore and maple trees (Aceraceae)
   pollen, *114*
Acer saccharum, sugar maple (North America), **151**
acquired characters, 13
adaptation, 24, 36, 154, 189
Adenostoma, chaparral shrubs (Rosaceae, California), 109
Aepyornithiformes, *see* elephant birds
Africa
   east, 56, 62
   glaciations, **88**
   mammal extinctions, 179
   northwest, 72
   vegetation change
      Holocene, 127–131
      Quaternary, 126–127
Agathis, gymnosperm trees (Araucariaceae, Australasia), **151**
aims, 2–3
albedo, 51, 85

Alces alces, elk, moose (Cervidae, Eurasia and North America), **156**, 157
Alces gallicus, extinct elk (Cervidae), **156**
Alces latifrons, extinct elk (Cervidae), **156**
Alchemilla, herbs (Rosaceae)
   pollen, *128*
Alchornea, tropical shrubs (Euphorbiaceae), 127
   pollen, *128*
alder, *see Alnus*
allelic diversity, 167
Alnus, alder trees (Betulaceae), 96, 100, 111, 126
   pollen, *97, 109, 110, 125*
Alnus glutinosa, alder (Europe), 107, **151**
Alnus jorullensis, aliso (South America)
   pollen, *127*
Alnus rubra, red alder (North America), **151**
Alopex lagopus, arctic fox (Canidae), *144*
Alternanthera, tropical herbs (Amaranthaceae), 122
   pollen, *123*
Amaranthaceae, herbs and shrubs (dicot angiosperms)
Ambrosia dumosa, ragweed herbs (Compositae), 117
ammonites
   longevity, **175**
Anemia, tropical ferns (Schizaeaceae)
   pollen, *127*
Animal Ecology, 36
animal ecology, 36
Antarctica
   glaciations, **88**
aphelion, 48, 56
Aphodius, beetles (Scarabaeidae)
Aphodius bonvouloiri, *139*
Aphodius holderi, 138
Apodemus sylvaticus, wood mouse (Muridae, Europe), *144*

Araliaceae, trees, shrubs, and climbers (dicot angiosperms), 127
pollen, *128*
Araucaria, gymnosperm trees (Araucariaceae), 132
pollen, *131*
Araucariaceae, coniferous trees (gymnosperms), 136
pollen, *136*
Arbutus menziesii, Pacific madrone (Ericaceae, North America), 109
Archaean, **4**, **88**
Arctica islandica, bivalve molluscs, 147
Arctostaphylos, shrubs (Ericaceae), 109
Artemisia, herbs and shrubs (Compositae), 96, 99, 100, *122*
pollen, *75*, *97–99*, *101*, *110*
Artemisia sec. *tridentatae*, sage brush (North America), *120*
Artemisia tridentata, sage brush (North America), 117
Asaphidion yukonense, beetles (Carabidae), 139
ash, *see Fraxinus*
Asia
glaciations, **88**
southeast, 56, 62
vegetation change
Quaternary, 136–137
Asteroideae, *see* Compositae (Tubuliflorae)
Atriplex confertifolia, shadscale (Chenopodiaceae, North America), 117, *120*
macrofossils, *118*
Australasia
vegetation change
Quaternary, 131–136

Bahamas, 162
Balanops, evergreen trees (Balanopaceae, southwest Pacific), **151**
Bandung basin, Java, 137
basswood, *see Tilia*
bear, polar, *see Ursus maritimus*
beech, *see Fagus*
southern, *see Nothofagus*
beetles
distribution change, 137–139
evolution, 157–159
extinction, 179
longevity, **175**

Bellamya unicolor, prosobranch molluscs, *163*
Beloperone, tropical herbs and shrubs (Acanthaceae)
Bermuda
Quaternary geology, 161
Betula, birch trees (Betulaceae), 100, 104, **151**, *169*
pollen, *109*, *124*
biogeographers
history-orientated, 38
bioturbation, 67
birch, *see Betula*
birds
extinction, 179–181
blackbrush, *see Coleogyne ramosissima*
Blepharis, shrubs (Acanthaceae)
pollen, *130*
Bogotá, Colombia, *125*, 124–126, *127*
Bovidae, African, 32
BP (before present), 3
brittle bush, *see Encelia farinosa*
Brunhes Chron, 133
Brunhes–Matuyama magnetic reversal, 67, 97, 133
bryophytes
longevity, **175**
Bufo terrestris, southern toad (Bufonidae, North America), *142*
Bulinus truncatus, pulmonate molluscs, *163*
Byrsonima, tropical trees (Malpighiaceae, Americas), 122
pollen, *123*

Caelatura, bivalve molluscs
Caelatura bakeri, *163*
Caelatura monceti, *163*
Caesalpinia bonduc, shrub (Leguminosae, Easter Island), 182
calcite, 65
calcium carbonate variations, 72, 74, *74*, *76*
calcium sulphate variations, 82, *84*
Calosoma reticulatum, beetles (Carabidae), 138, *139*
Camarhynchus, finches (Fringillidae, Galápagos)
Camarhynchus crassirostris, *170*
Camarhynchus heliobates, *170*
Camarhynchus pallidus, *170*
Camarhynchus parvulus, *170*
Camarhynchus pauper, *170*
Camarhynchus psittacula, *170*

Index

Cambrian, 4, **88**
Capparis, shrubs and trees (Capparidaceae), 129
Capparis decidua, 128
Capreolus capreolus, roe deer (Cervidae), 145, **156**
carbon dioxide, atmospheric, 51, 89
carbonate-dissolution oscillations, 72
Carboniferous, 4, 82, *83*, **88**, 197
Carphoborus andersoni, beetles (Scolytidae), 139, *140*
Carpinus, hornbeam trees (Betulaceae), *see also Carpinus/Ostrya*, 100, **152**, *169*
  pollen, *98, 99, 124*
Carpinus betulus, hornbeam (Betulaceae, Europe), 100–102
  pollen, *97, 98, 101*
Carpinus/Ostrya, *see also Carpinus* and *Ostrya*, **151**
  pollen, *97, 101*
Carya, hickory trees (Juglandaceae), 100, 111, 181
  pollen, *114*
Carychium bermudense, pulmonate molluscs, 162
Caryophyllaceae, mostly herbs (dicot angiosperms), 127
  pollen, *97, 128*
Castanea, chestnut trees (Fagaceae), 100, 111
Castanea dentata, American chestnut (Fagaceae, North America)
  pollen, *114*
Castanopsis, 'chestnut' trees (Fagaceae)
  pollen, *136*
Castela emoryi, trees and shrubs (Simaroubaceae), 117
Casuarina, sclerophyll trees (Casuarinaceae), 132, 133, 181
  pollen, *131, 134*
Cathormiocerus, beetles (Curculionidae)
Cathormiocerus curviscapus, 138, *139*
Cathormiocerus validiscapus, *139*
CCM, 51, 56, 61, 62, 87
Ceanothus, shrubs (Rhamnaceae), 109, 111
Cecropia, tropical pioneer trees (Cecropiaceae), 122
  pollen, *123, 127*
cedar, *see Cedrus*
Cedrela, tropical American trees (Meliaceae)
  pollen, *127*

Cedrus, cedar trees (Pinaceae), 100
Celtis, trees (Ulmaceae), 100
Cenozoic, 4, 66, 72, 85, 87, 88, 92, 126
Central America
  vegetation change
    Quaternary, 122–124
Cerion, pulmonate molluscs
Cerion agassizi, *162*
Cerion spp., 162–163
Certhidea olivacea, finch, Galápagos (Fringillidae), *170*
Cervus elaphus, red deer (Cervidae), 145, 155, *155*, **156**
Cervus sp., *156*
chapparal, 111
charcoal, *131*, 132, 133, *134*
Chenopodiaceae, herbs and shrubs (dicot angiosperms), 96, 99, 100, 122
  pollen, *75, 97, 101, 123, 134*
Chenopodiaceae–Amaranthaceae, 130
  pollen, *123, 124*
Chenopodiineae, *see* Chenopodiaceae
chestnut, *see Castanea*
chipmunk
  eastern, *see Tamias striatus*
Chlamys, bivalve molluscs
Chlamys flexuosa, 148
Chlamys glabra, 148
chronology, 41, 66, 71
Chrysolepis, 'chestnut' trees (Fagaceae), 111
  pollen, *110*
Cichorioidideae, *see* Compositae (Liguliflorae)
Cionichthys, redfieldiids (chondrostean fishes), *80*
classes, 30, 185
Clear Lake, California, USA, *110*, 109–111
Clemmys insculpta, wood turtle (Emydidae, North America), *142*
Cleome, tropical herbs (Capparidaceae)
  pollen, *127*
Cleopatra ferruginea, prosobranch molluscs, *163*
Cliffortia, tropical shrubs (Rosaceae, Africa), 127
Cliffortia nitidula
  pollen, *128*
climate
  Galápagos, 171
  glacial, 123
  interglacial, 101

Index · 229

models, 51–64, 87–89
  Cretaceous, 61–62
  Holocene, 51–56
  Permian, 56
  modern, *52, 57*
  reconstruction
    18 ka, *52*
    9 ka, *57*
    Cretaceous, *63, 64*
    Pangaea, *61*
climatic change, 185, 188, 191, 196
  and vegetation change, 42
  astronomical forcing, 185
  astronomical-forcing, 9, 10, 90
  biotic response, 8, *32, 33*, 111, 113, 121, 124, 127, 131, 135, 138, 139, 144, 145, 147, 148, 150, 152, 154, 158, 160, 175, 183–184, 191, 194, 197
  cause, 2, 65
  evolutionary change, 24, 31, 35, 43, 154
  frequency, 191
  Holocene, 56, 140
  Permian, 82
  pre-Quaternary, 185
  Quaternary, 22, 24, 32, 40, 43, 66, 70, 71, 145, 184
  speciation, 194
  throughout Earth history, 185
  Triassic, 81
  tropical, 56, 131, 185
climatic oscillations, 78, 94, 125, 131, 132, 164, 174, 176, 185, 189, 190
  biotic response, 3, 197
  Cretaceous, 62
  pre-Quaternary, 47, 62, 185
  Quaternary, 18
climatic record
  independence, 51
coccolithophores, minute calcareous marine planktonic algae, 146, *146*, 147
  distribution change, *146*
Coccolithus pelagicus, coccolithophores, *146*, 147
Coleogyne ramosissima, blackbrush shrubs (Rosaceae), *120*
Commiphora, trees and shrubs (Burseraceae), 129
  pollen, *130*
communities
  as classes, *see* classes
  concepts, 37

development, 95, 103, 116–119, *120*
  disruption, 185
  recurrence, 96, 195–196
  stability, 150
  temporary, 103, 113, 117, 142, 148, 150, 185
Community Climate Model, *see* CCM
competition, 18, 29, 36, 183, 185, 187, 195
Compositae, herbs and shrubs (dicot angiosperms)
  (Liguliflorae)
    pollen, *97*
  (Tubuliflorae)
    pollen, *75, 97, 123, 134, 182*
  pollen, *128*
Coprosma, trees and shrubs (Rubiaceae), 182
  pollen, *182*
corals
  distribution change, 148
  evolution, 164
Corbicula consobrina, bivalve molluscs, *163*
Coriaria ruscifolia, trees and shrubs (Coriariaceae)
  pollen, *127*
Cornus, trees and shrubs (Cornaceae)
  pollen, *124*
Corylus, hazel trees (Betulaceae), 100, 104, **152**
  pollen, *97, 98, 101, 109*
Corylus avellana, hazel (Europe), **151**
creosote bush, *see Larrea divaricata*
Cretaceous, **4**, 61, 62, *63*, 72, 74, 76, *76, 77*, 78, **88**, 92, 126, 181, 198
Croll, James, 10, 14, 41
Croton, trees, shrubs, and herbs (Euphorbiaceae, tropical)
  pollen, *127*
Croton bogotense
  pollen, *127*
Cruciferae, herbs (dicot angiosperms), 127
  pollen, *128*
Cryptomeria japonica, Japanese cedar (Taxodiaceae, east Asia), 137
Cunoniaceae, southern hemisphere trees and shrubs (monocot angiosperms), 132, **151**
  pollen, *131, 136*
Cupressaceae, coniferous trees (gymnosperms), 74
  pollen, *75*

Cupressus arizonica, Arizona cypress (Cupressaceae, North America), 117
Cyathea, tree ferns (Cyatheaceae), 122
  spores, *134*, *136*
Cycladophora davisiana, radiolaria, 67, 70, 147
cycles, 5
Cyclococcolithus leptoporus, coccolithophores, 146, 147
Cymindis unicolor, beetles (Carabidae), *140*
Cynomys ludovicianus, blacktail prairie dog (Sciuridae), *143*
Cyperaceae, sedges (monocot angiosperms), 128
  pollen, *75*, *130*

$\delta^{18}O$ (oxygen isotopic ratios), 5, 65, 67, 69–73, 75, 94, 109, 125, 132, 133, 155, 185
Dacrycarpus, gymnosperm trees (Podocarpaceae), 136
  pollen, *136*
Dacrydium, gymnosperm trees and small shrubs (Podocarpaceae), 132, 133, 181
  pollen, *131*
Dama dama, fallow deer (Cervidae), 145, **156**
Daphnopsis, tropical American trees (Thymelaeaceae)
  pollen, *125*
Darwin, Charles, 6, 9–15, 18, 22, 25, 35, 36, 40, 42, 169, 187
Dasycercus cristicauda, mulgara, marsupial (Dasyuridae, Australia), *144*
de Vries, Hugo, 14–15
Deep Sea Drilling Project, 72
deer
  distribution change, 144–145
  evolution, *155*, **156**, 155–156
  fallow, *see Dama dama*
  red, *see Cervus elaphus*
  roe, *see Capreolus capreolus*
Denticulopsis seminae, marine diatoms, 147
descent with modification, 12
desert vegetation, 100
desman
  Russian, *see Desmana moschata*
Desmana moschata, Russian desman (Talpidae), *144*
Devils Hole, Nevada, 65
Devonian, **4**, 82, **88**

diatoms, golden-brown unicellular algae, 147, *167*
  marine
    longevity, **175**
Dicliptera, tropical herbs (Acanthaceae)
Dicliptera–Beloperone
  pollen, *127*
Dicrostonyx hudsonius, Hudson Bay collared lemming (Arvicolidae, North America), *143*
Dicrostonyx torquatus, collared lemming (Arvicolidae, North America), *143*, *144*
Dinornithiformes, *see* moas
Diplurus, coelacanths (sarcopterygian fishes), *80*, *81*
Diplurus longicaudatus, *81*
Discus ruderatus, gastropod molluscs, *141*
dissimilarity mapping, 103, 115, *116*
distribution change, 2, 3, 7, 33, 92–154, 194
  altitude, *118*
  beetle subspecies, 158
  cause, 8, 22
  European trees, *169*
  frequency, 184
  genetic consequences, 168, 193
  interglacial, 168
  late-Quaternary
    deer, 156
    European *Quercus*, *104*
    European *Tilia*, *105*
  process, 193
  speciation, 176
  species, 150
  time-scales, 42
disturbance, pollen indicators, *136*
DNA, 144
Dobzhansky, Theodosius, 16, 25
Dodonaea viscosa, tropical trees and shrubs (Sapindaceae)
  pollen, *127*
Drimys, evergreen trees and shrubs (Winteraceae)
  pollen, *125*, *134*
Drimys granadensis
  pollen, *127*
dwarfing, 155

Earth, age, 40, 41
Earth, orbital parameters, *see* orbital parameters, Earth

Index · 231

Earth–Moon distance, **47**
Easter Island, 182
extinction, plant, 181–183
Echinocactus horizonthalonius, cacti (Cactaceae), 117
ecological interactions, 13
ecological moments, 29, 190
ecological process, 35–39
ecological theory, 43
ecologists, 1, 150
  and Quaternary time-scales, 41
  machinery-orientated, 38
ecology, 2
  and evolution, 3
  and evolutionary change, 36, 38
  and palaeontology, 3
  definition, 35, 36
  evolutionary, 38
  evolutionary change, 37
  problems, 36
  scope, 36
El Valle, Panama, *123*, 122–123
Elaeocarpus, tropical trees (Elaeocarpaceae), 132, **151**
  pollen, *131*
Eldredge, Niles, 26
elephant birds, extinct Madagascan birds (Aepyornithiformes), 180
elm, *see Ulmus*
Encelia farinosa, brittle bush (Compositae), *120*
environmental change, 90, 175, 176, 183, 185, 187, 191
  and evolutionary change, 43
  biotic response, 8, 16, 24, *32*, *33*, 103, 186
  evolutionary change, 24, 31
  speciation, 176
environmental tracking, 196–197
Eocene, **4**, 72, **88**
Ephedra, evergreen, shrubby switch plants (gymnosperms, Ephedraceae), 74
  pollen, *75*
Ephedra viridis, 117
equinox, *49*
Ericaceae, trees and shrubs (dicot angiosperms), 127, 128
Eriocaulon, mostly tropical herbs (Eriocaulaceae)
Eriosema, herbs (Leguminosae)
Eriosema (cf.)

pollen, *127*
Eucalyptus, Australasian sclerophyll trees (Myrtaceae), 131–133, 181
  pollen, *134*
Eucalyptus-type
  pollen, *131*
Eucladoceros, comb-antlered deer, extinct (Cervideae)
Eucladoceros ctenoides, **156**
Eucladoceros falconeri, **156**
Eucladoceros sedgwicki, **156**
Eucladoceros tetraceros, **156**
Eucommia, tree, known only from east Asia today (Eucommiaceae), 100, 181
Eugenia, tropical trees (Myrtaceae), **151**
Eugenia-type
  pollen, *136*
Eupera ferruginea, bivalve molluscs, *163*
Europe
  extinction, plants, 181
  glaciations, **88**
  vegetation change
    Holocene, 102–107
    Quaternary, 96–102
evolution, *see* evolutionary change
  and ecology, 3
  gradual, 175
Evolution: the Modern Synthesis, 22
evolutionary change, 2, 3, 29, *32*, 154–177, 195
  above species level, 24
  and ecology, 36, 38
  and environmental change, 43
  climatic change, 24, 31, 35, 43, 154
  constraint, 186
  ecology, 37
  environmental change, 24, 31
  equilibrium, 186
  fact of, 35, 38
  frequency, 184
  gradual, *15*, 32
  Holocene, 173
  modes, 20, **23**
  phyletic, 22, **23**
  plants, 22, 24
  process, 26, 38
  quality of evidence, 154
  quantum, 22, **23**
  Quaternary, 173
  random, 186
  rates, 20, 24, 34, 40, 43

selection, 186
theory of, 10
evolutionary ecology, 38
evolutionary hierarchy, 29, **189**, 189–190
evolutionary progress, 19, 29, 38, 190
evolutionary synthesis
  modern, *see* modern synthesis
  post-modern, 184–195
evolutionary theory, 43
exaptation, 189
extinction, 2, 3, 7, 21, *32*, 33, 154, 178–183, 188
  birds, 180–181
  cause, 31
  frequency, 183, 184
  mammals, *179, 180*
  mass, *see* mass extinctions
  plants, 96, 100, 117, 119, 132, 133, 137
  rates, 8, 178
extrapolation, 19, 26, 33, 35, 38, 154

Fagus, beech trees (Fagaceae), 100–102, 111, 114, **151, 152**
  pollen, *97, 101, 109*
Fagus grandifolia, American beech (Fagaceae, North America), **151**
  pollen, *114*
Fallugia paradoxa, shrubs (Rosaceae), 117
  macrofossils, *118*
fauna
  disharmonious, 141
  intermingled, 142
    beetles, European, *139*
    beetles, North American, *140*
    mammals, Australian, *144*
    mammals, Eurasian, *144*
    mammals, North American, *143*
    molluscs, European, *141*
    vertebrates, North American, *142*
FAUNMAP, 141
Ficalhoa, tropical African shrubs (Theaceae), 127
  pollen, *128*
Filicales, ferns (Filicopsida)
  spores, *182*
finches, Darwin's, 169, 170, *170*, 171, 172, *172*, 173, 175–177
fir, *see Abies*
  Douglas, *see Pseudotsuga*
fishes
  evolution, 157
  fossil, 79, *80, 81*
freshwater
  longevity, **175**
foraminifera, testate protozoans, 147
  benthic
    longevity, **175**
  planktonic
    longevity, **175**
forest
  development, 104, 111, 115
  Holocene, 42, 123, 127, 137
  Holocene spread, 42, 103, *109*, 111, 112, *112–114*
  interglacial, 96, 99, 100, 102, 122, 125, 133, 135
  modern, 103, 111
  Quaternary, 126
  rates of spread, 114
fossil record, 20, 27, 34, 35, 43
  ignored, 34
  interpretation, 26, 31
  Quaternary, 194
  use, 34, 38
founder effect, 168
fox, arctic, *see Alopex lagopus*
Fraxinus, ash trees (Oleaceae)
  pollen, *109, 124*
Fraxinus anomala, single-leaved ash (Oleaceae, North America), *120*
Fraxinus excelsior, ash (Oleaceae, Europe), 100
  pollen, *98*
Fraxinus nigra, black ash (Oleaceae, North America), **151**
future change, 185

Gabbiella senaariensis, prosobranch molluscs, 163
Galápagos, 169, 171, 173, 176
Gastrocopta rupicola, gastropod molluscs, 140
GCM, 51, *52, 57*, 89
general circulation model, *see* GCM
generation times, 20, 174, 176
geneticists, 35
genetics, 1, 3, 14–16, 19, 29, 36
Genetics and the Origin of Species, 16
geological record
  imperfect, 13
geologists
  and Quaternary time-scales, 41
Geospiza, finches (Fringillidae, Galápagos)

Geospiza conirostris, 170, 172
Geospiza difficilis, 170
Geospiza fortis, 170, 171, 172
Geospiza fuliginosa, 170
Geospiza magnirostris, 170
Geospiza scandens, 170
Gephyrocapea caribbeanica, coccolithophores, 146, 147
Gilia, herbs (Polemoniaceae, North America)
Gilia-type
 pollen, *75*
glacial, 94, 132
glacial period, 9, 13, 15, 40, 41, 85, 89, 96, 122, 135
 species altered, 14, 21
 species persistence, 9, 15, 15
glacial, last, 40, 51, 102, 103, 117, 123, 135, 137, 138, 146, 162, 178, 181
 definition, 5
glacial–Holocene transition, 20, 41, 94, 178, 179, 181
glacial–interglacial oscillations, 40, 95, 101, 122, 133, 137, 138, 150, 156, 161, 163, 191, 193
glaciations, 1, 2, 14, 18, 24, 39, *93*, 92–94, 185, 197
 and species formation, 18
 causes, 31, 85–89
 north–south contrasts, 14, 70
 pre-Quaternary, 10, **88**
 Quaternary, 66
Globigerina bulloides, foraminifera, 67
Globoconella, foraminifera, 165, *165*
Globoconella conomiozea terminalis, 165, *165*
Globoconella pliozea, 165, *165*
Globoconella sphericomiozea, 165
Gould, John, 169
Gould, Stephen Jay, 26
Gramineae, grasses (monocot angiosperms), 96, 99, 100, 127, 128
 pollen, *75*, *97–99*, *101*, *110*, *123*, *125*, *128*, *130*, *134*, *182*
Grande Pile, France, 102
graptolites
 longevity, **175**
grasses, *see* Gramineae
grassland, 135
Grewia tenax, tree (Tiliaceae, Africa), 128, 129

Haeckel, Ernst, 35

Hagenia abyssinica, east African shrub (Rosaceae), 127
 pollen, *128*
hazel, *see Corylus*
Hedyosmum, tropical plants (Chloranthaceae)
 pollen, *123–125*
Helicospaera carteri, coccolithophores, *146*
Helophorus, beetles (Hydrophilidae)
Helophorus aquaticus, 158, *158*
Helophorus arcticus, 140
Helophorus aspericollis, 138
Helophorus brevipalpis, 138
hemlock, *see Tsuga*
herbs
 pollen, *see also* individual taxa, *75*, *123*, *124*
heredity, 14
Hermanites transoceanica, marine ostracodes (Crustacea), 160
heterozygosity, 24
Histiopteris, tropical ferns (Dennstaedtiaceae)
Histiopteris incisa-type
 spores, *127*
Holocene, **4**, *71*
 climate, 41
 forest, *see* forest, Holocene
 status, 5
 usage, 4
 vegetation change, *see* vegetation change, Holocene
Hooker, Joseph, 9
hornbeam, *see Carpinus*
hop, *see Ostrya*
Hura, tropical trees (Euphorbiaceae), 122
Huxley, Julian, 25, 187
hybridization, 24
Hymenophyllum, filmy ferns (Hymenophyllaceae)
Hymenophyllum myriocarpum-type
 spores, *127*
Hypericum, trees, shrubs and herbs (Guttiferae)
 pollen, *125*

ice-ages, *see* glaciations
ice-sheets, continental, 4, 50, 51, 56, 67, 70, 85, 183, 185
Ilex, holly trees and shrubs (Aquifoliaceae)
 pollen, *124*, *136*

individualistic behaviour, 37, 42, 103, 113, 121, 139, 150, 185, 188, 194, 196
individuals, 30–31, 150, 185, 196
inheritance
  blending, 16
  particulate, 16
insolation, 44, *50*, 48–51, 61, *69*, 124
interglacial, 40, 41, 94, 99, 132, 135, 147
  distribution change, 168
  last, 40, *71*, 101, 102, 110, 122, 155, 158, 162
  definition, 5
  present, *see* Holocene
interstadial, 94
Ioannina, Greece, *101*, 100–101
isolation, 19, 20, 22, 24, 27, 160, 173
  geographic, 15, 18, 27, 164, 171, 189, 191
  reproductive, 26, 32

Jubaea chilensis, Chilean wine-palm (Palmae), 182
Juglans, walnut trees (Juglandaceae) pollen, *124, 125*
Juniperus, juniper shrubs (Cupressaceae), 116, *119, 120,* **152**
  pollen, *97*, 124
Juniperus osteosperma, Utah juniper (Cupressaceae, North America), 117, *122*
  macrofossils, *118*
Juniperus osteosperma/monosperma, Utah and one-seed junipers (Cupressaceae, North America), *122*
Juniperus scopulorum, Rocky Mountain juniper (Cupressaceae, North America), 117, 119, *122*
Jupiter, 45
Jurassic, **4**, 78, **88**, 157, 191

Kashiru, Burundi, 126–127, *128*
koala, *see Phascolarctos cinereus*
Kochia, desert shrubs (Chenopodiaceae), 99
Krascheninnikovia, desert shrubs (Chenopodiaceae), 99

Lake Biwa, Japan, 137
Lake George, Australia, *134*, 133–135
Lake Hordorli, New Guinea, 135, *136*
Lake Quexil, Guatemala, *124*, 123–124
larch, *see Larix*
Larix, larch trees (Pinaceae), 100

Larix gmelinii, Dahurian larch (Pinaceae, Asia), 137
Larix laricina, tamarack (Pinaceae, North America), 111
  pollen, *112*
Larrea divaricata, creosote bush (Zygophyllaceae, North America), 117, *120*
Lathyrus, herbs (peas) (Leguminosae) pollen, *127*
Lauria cylindracea, gastropod molluscs, *141*
lemming
  collared, *see Dicrostonyx torquatus*
  Hudson Bay collared, *see Dicrostonyx hudsonius*
  northern bog, *see Synaptomys borealis*
  Norway, *see Lemmus lemmus*
Lemmus lemmus, Norway lemming (Arvicolidae), *144*
Les Echets, France, 102
lime, *see Tilia*
limnology, 36
Linnean Society, 10, 11
Liomys irroratus, Mexican pocket mouse (Heteromyidae), *143*
Liquidambar, deciduous trees (Hamamelidaceae), 100, 181
Lycium carolinianum, shrub (Solanaceae, Easter Island), 182
Lycopodium, club-mosses (Lycopodiaceae) spores, *134*
Lyell, Charles, 6–10, 15, 35, 39, 40, 42, 43
  letter to Darwin, 9
  views on species, 7–9
Lynch's Crater, Australia, *131*, 131–133
Lysipomia, Andean herbs (Campanulaceae) pollen, *127*

Macaranga, tropical trees and shrubs (Euphorbieaceae), 126, 127
  pollen, *128*
macroevolution, 13, 16, 19, 26, 29, 35, 188, 191
Madagascar, 180
  extinction, mammals, 181
Maerua, shrubs and trees (Capparidaceae) pollen, *130*
Maerua crassifolia, 128
mammals
  evolution, 155–157
  extinction, 178–181

longevity, **175**
marine microfossils
  distribution change, 145–147
mass extinctions, 29–30, 72, 176, 189, **189**, 190, 191, 194
mathematical ecology, 36
Mayr, Ernst, 18–20, 24, 25, 27, 43, 164, 187
mega-evolution, 22
Megaceros, giant deer, extinct (Cervideae)
Megaceros dawkinsi, **156**
Megaceros giganteus, **156**
Megaceros savini, **156**
Megaceros verticornis, **156**
Melanoides tuberculata, prosobranch molluscs, *163*
Melastomataceae, herbs and shrubs (dicot angiosperms), 122
  pollen, *123–125*
Meliaceae, tropical trees (dicot angiosperms)
  pollen, *123*
Mendel, Gregor, 14, 16
Mesozoic, **4**, 157
Miconia, trees and shrubs (Melastomataceae)
  pollen, *125*
Microcachrys, gymnosperm trees (Podocarpaceae, Tasmania), 133, 181
microevolution, 26, 29, 35, 37, 190
Microstrobos, gymnosperm trees (Podocarpaceae, Australia), 181
Microtus pennsylvanicus, meadow vole (Cricetidae, North America), 157
Milankovitch hypothesis, 65
Milankovitch oscillations, 48
Milankovitch, Milutin, 48
Miocene, **4**, 72, 73, 93
moas, extinct New Zealand birds (Dinornithiformes), 180
modern synthesis, 15–26, 34–35, 42, 173, 191
molecular biology, 29
molluscs
  marine, 148
    bivalves, longevity, **175**
    bivalves, Lyellian percentages, *174*
    distribution change, 147–148
    gastropods, longevity, **175**
  non-marine
    distribution change, 139–140
    evolution, *163*, 161–164

Tertiary, 9
monsoons, 56, 129–131
Moon, 44
Moraceae, herbs, shrubs, and trees (dicot angiosperms), 123
Mougeotia, trees and shrubs (Sterculiaceae)
Mougeotia laetevirens-type
  pollen, *127*
mouse
  grasshopper, *see Onchyomys leucogaster*
  Mexican pocket, *see Liomys irroratus*
  wood, *see Apodemus sylvaticus*
Muelenbeckia, climbers and creepers (Polygonaceae)
  pollen, *127*
mulgara, *see Dasycercus cinereus*
mutation, 14, 15, 20, 26, 28, 186
Mutela nilotica, bivalve molluscs, *163*
Myrica, shrubs (Myricaceae)
  pollen, *123*, *125*
Myrtaceae, trees and shrubs (dicot angiosperms), 122, 132
  pollen, *123*, *125*, *134*
Myrtaceae (*Eugenia*-group)
  pollen, *131*

natural selection, *see* selection, natural
Neotoma, packrats and woodrats (Cricetidae), 115, *118*, *119*, 121
Neotoma floridana, 143
New Zealand, 181
  extinction, plants, 181
North America, 62, 87
  glaciations, **88**
  vegetation change
    Holocene, 111–115
    Quaternary, 108–121
Nothofagus, southern beech trees (Fagaceae), 132, 133, 135, 136, 181
  pollen, *136*
Nothofagus brassii-type
  pollen, *134*

oak, *see Quercus*
oceanography, 36
Oenothera, evening primroses (Onagraceae), 14
Okhotsk, Sea of, 147
Olea, olive trees (Oleaceae), 96, 102, 127, **152**
  pollen, *97*, *128*
Oligocene, **4**, 72, **88**, 93, 126

olive, *see Olea*
On the Origin of Species, 1, 9, 11, 14, 16, 19, 36, 39, 40
Onagraceae, herbs and shrubs (dicot angiosperms)
  pollen, *124*
Onchyomys leucogaster, grasshopper mouse (Cricetidae, North America), 144
Onthophagus gibbulus, beetles (Scarabaeidae), *139*
Opetiopalpus scutellaris, beetles (Cleridae), *139*
Opisthius richardsoni, beetles (Carabidae), *140*
orbital change hypothesis, 66
orbital parameters
  Earth, 9
orbital parameters, Earth, 44–51, 96, 125
  eccentricity, 45, *46*, 48–50, 56, 62, 70–72, 75, 82, 84
  obliquity, 45, 46, *46*, *47*, 48, 50, 56, 61, 62, 70, 71, 78, 81, 82, 84
  precession, 45, 46, *46*, *47*, 48, *49*, 50, 56, 62, 70, 78, 82, 84, 96
orbital parameters, Mars, *48*
orbital variations, 10, 14, 47, 48, 51, 62, 67, *69*, 71, 90, 132, 185, 187
  Cretaceous, 76
Ordovician, **4**, 82, *87*, **88**, 197
ostracodes, Crustacea
  evolution, 159–160
Ostrya, hop hornbeam trees (Betulaceae), *see also Carpinus/Ostrya*
  pollen, *98*, *124*
overkill, 181
oxygen isotopic ratios, *see* $\delta^{18}O$

packrat, *see Neotoma*
packrat middens, 115–121
Paepalanthus, tropical herbs (Eriocaulaceae)
Paepalanthus–Eriocaulon
  pollen, *127*
palaeomagnetism, 73, 133
palaeontologists, 1
  history-orientated, 38
palaeontology, 1, 15
  and ecology, 3
Paleocene, **4**
Paleozoic, **4**
Palmae, palms (monocot angiosperms)
  pollen, *182*

Panama, Isthmus of
  formation, 126, 159
Pangaea, 56, *61*
paradox of the first tier, 29, 30
Parrotia, trees (Hamamelidaceae), 100
peacock's tail, 192
Pecten jacobaeus, bivalve molluscs, 148
periglacial processes, 94
perihelion, 14, 45, 48, 56
Permian, **4**, 56, 82, *84*, **88**
Perspectives in Ecological Theory, 1
Phanerozoic, **4**, 197
Phascolarctos cinereus, koala (Phascolarctidae, Australia), *144*
phyletic gradualism, 27, 34, 43, 154, 166
Phyllocladus, gymnosperm trees (Phyllocladaceae), 132, 133, 136, 181
  pollen, *134*, *136*
Picea, spruce trees (Pinaceae), 102, 111, *120*, **151**
  pollen, *75*, *110*, *112*
Picea engelmannii, Engelmann spruce (Pinaceae, North America), *122*
Picea glauca, white spruce (Pinaceae, North America), 115, **151**
Picea pungens, Colorado spruce (Pinaceae, North America), *122*
Piliostigma, usually climbers (Leguminosae), 128, 129
Pinaceae, coniferous trees (gymnosperms)
Pinaroloxias inornata, Cocos Island finch (Fringillidae), *170*
pine, *see Pinus*
Pinus, pine trees (Pinaceae), 99, 100, 110, 111, *119*, *120*, **152**, *169*
  pollen, *75*, *97–99*, *101*, *109*, *110*, *124*
Pinus banksiana/resinosa, jack pine/red pine (Pinaceae, North America), 111
  pollen, *112*
Pinus contorta, lodgepole pine (Pinaceae, North America), **151**, 167
Pinus edulis, pinyon (Pinaceae, North America), *122*
Pinus flexilis, limber pine (Pinaceae, North America), 117, 119, *120*, *122*
  macrofossils, *118*
Pinus longaeva, Great Basin bristlecone pine (Pinaceae, North America)
  macrofossils, *118*
Pinus monophylla, single-leaf pinyon (Pinaceae, North America), 117

Index · 237

macrofossils, *118*
Pinus ponderosa, Ponderosa pine (Pinaceae, North America), 119, 120, *120, 122*
macrofossils, *118*
Pinus remota, papershell pinyon (Pinaceae, North America), 116
Pinus sabiniana, digger pine (Pinaceae, North America), 109
Pinus strobus, white pine (Pinaceae, North America), **151**
pollen, *113*
Pinus sylvestris, Scots pine (Pinaceae, western Eurasia), **151**
plant ecology, 36
plants
   distribution change, 95–137
   evolution, 167–168
   evolutionary change, 22, 24
   extinction, 96, 100, 117, 119, 132, 133, 137, 181–183
   longevity, **175**
plate tectonics, 85, 187
Pleiodon, bivalve molluscs, *163*
Pleistocene, 4, **4**
Pliocene, **4**, 93
Poaceae, *see* Gramineae
Podocarpaceae, coniferous trees (gymnosperms), 133
Podocarpus, gymnosperm trees (Podocarpaceae), 126, 127, 132, 133, **151**
pollen, *125, 128, 131, 134, 136*
Poecilozonites, pulmonate molluscs
Poecilozonites bermudensis, 161
Poecilozonites spp., 161–162
pollen
   analysis, 41
      resolution, 95
   stratigraphy, 41
pollen data
   Holocene
      Africa, 127–131
      Europe, 102–107
      North America, 111–115, 167–168
      Sahara, *129*
   Miocene, 73–75
   proxy for past climate, 42
   Quaternary
      Africa, 126–127
      Asia, 136, 137
      Australasia, 131–136

      Central America, 122–124
      Easter Island, 181–183
      Europe, 96–102
      North America, 109–111
      South America, 124–126
Polynesian colonization, 135, 180, 182
polyploidy, 24, 192
population
   density, 183
   ecology, 37
   global change, 147
   increase, 10, **21**, 25, 36, 101, 102, 107, 115, 150, **151, 152**, 153, 154, **187**, 188
   interaction, 190
   persistence, 168
   size, 16, 20, 36, 145, 151, 176, 186
   splitting, 22
   stability, 11, **21**, 25, **187**, 194
   structure, 24, 183
population-thinking, 18
prairie dog
   blacktail, *see Cynomys ludovicianus*
preadaptation, 22, 24, 189
Principles of Geology, 7, 9, 10
Priscoan, **4**
Proterozoic, **4**, 84, 85, *87*, **88**
Protista
   evolution, 165–167
Pseudobovaria, bivalve molluscs, *163*
Pseudotsuga, Douglas fir trees (Pinaceae), 111, *122*
   pollen, *110*
Pseudotsuga menziesii, Douglas fir (Pinaceae, North America), 109, *120*, **151**
Pteris, ferns (Adiantaceae)
Pteris grandifolia-type
   spores, *127*
Pterocanium, radiolaria
Pterocanium charybdeum, *166*
Pterocanium prismatium, *166*
Pterocanium spp., 166
Pterocarya, wing nut trees (Juglandaceae), 96, 100
   pollen, *99*
Pterostichus punctatissimus, beetles (Carabidae), *140*
Punctuated Aggradational Cycle, 89
punctuated equilibria, 15, 19, 26–31, 31, 34–35, 43, 166, 167, 188, 190
Puriana, marine ostracodes (Crustacea)

Puriana aff. *elongorugata*, 159
Puriana carolinensis, 160
Puriana floridana, 159, *160*
Puriana mesacostalis, 159, *160*
Puriana minuta, 160
Puriana rugipunctata, 159

Quaternary, **4**
  chronology, 66
  chronostratigraphy, *71*
  climatic change, 22, 24, 32, 43, 66, 70, 71
  climatic oscillations, 18
  environmental dynamics, 184
  fossil record, 2
  glaciations, 66, 71
  history, 184
  history of attitudes, 2
  palaeoecologists, 1, 39, 42
  palaeoecology, 3, 41, 120
    ignored, 1
  paradox, 197
  representativity, 185, 197–198
  research, 39–42
  usage, 4
  vegetation change, *see* vegetation change, Quaternary
Quercus, oak trees, deciduous and evergreen (Fagaceae), 96, 99, 100, 103, 109–111, 122, 123, 126, **151**, *169*
  pollen, *98, 99, 101, 104, 109, 110, 113, 123–125*
Quercus (deciduous), **152**
  pollen, *97*
Quercus douglasii, blue oak (Fagaceae, North America), 109
Quercus gambelii, Gambel oak (Fagaceae, North America), 120
Quercus ilex-type, evergreen oaks (Fagaceae, Europe)
  pollen, *97*

radiocarbon ages, 41, 42, 67, 96, 97, 109, 115, 122, 132, 133, 148, 182
radiocarbon years
  calibration, 4
radiolaria, marine planktonic protozoans, 67, 147
radiometric ages, 125
Rangifer tarandus, reindeer, caribou (Cervidae, Eurasia and North America), 145, **156**

Ranunculaceae, mostly herbs (dicot angiosperms)
  pollen, *97*
Ranunculus, herbs, buttercups (Ranunculaceae), 127
  pollen, *128*
Rapanea, trees and shrubs (Myrsinaceae), **151**
  pollen, *124*
recombination, <u>15</u>, 26, 186
Red Queen hypothesis, 26
redox variations, 74, *76*
refugia, glacial, 42, 100, 103, 105, 123, 135, 168
reindeer, *see Rangifer tarandus*
reptiles
  fossil, 79, *80*
Rhabdosphaera stylifera, coccolithophores, *146*
Rhamnaceae, trees and shrubs (dicot angiosperms), 111
  pollen, *110*
Rhamnus, shrubs (Rhamnaceae), 111
Rhizosolenia, marine diatoms
Rhizosolenia bergonii, *167*
Rhizosolenia praebergonii, *167*
Rhizosolenia spp., 166–167
Ribes, low shrubs (Grossulariaceae)
Ribes cf. *velutinum*
  macrofossils, 118

sagebrush, *see Artemisia* sec. *tridentatae*
Sahara, lakes, 127, *129*
saiga, *see Saiga tatarica*
Saiga tatarica, saiga (Bovidae, Eurasia), *144*
Salix, willow trees, (Salicaceae), **151**
  pollen, *75*
Salvadora, trees and shrubs (Salvadoraceae), 128
  pollen, *130*
Samerberg, Germany, 102
Sapotaceae, trees and shrubs (dicot angiosperms)
  pollen, *123*
Sarcobatus, greasewood, shrub (Chenopodiaceae), 74
  pollen, *75*
Satureja, herbs (Labiatae)
Satureja nubigena-type
  pollen, *127*
savanna, 128, 129

sea-level change, 51, 90, 135, 148, *149*, 155, 162, 164, 175, 191
  Carboniferous, 82, *83*
  Proterozoic, 85, *87*
  Quaternary, 94, 148, 161
  Triassic, 78
sea-surface temperatures, 67
secondary contact hypothesis, 25
sedges, *see* Cyperaceae
sediment
  deep-sea, 5, 40, 65, 67, *68*, 71, 72
  laminated, 65, 73, 79, 82, 96, 157
  periodicity, 72, 73, 75, 78, 79, 82, 84, 85
sedimentary variations
  Cenozoic, 66–74
  Mesozoic, 74–82
  Paleozoic, 82–84
  Proterozoic, 84–85
Selaginella, club-mosses (Selaginellaceae)
Selaginella sellowii-type
  spores, *127*
selection, 20, 187
  artificial, 12
  natural, 9–14, 16, <u>16</u>, 20, **21**, 22, 25, <u>25</u>, 26, <u>28</u>, 35, 36, 38, <u>38</u>, 154, 168–169, 171–173, 175, 186–187, **187**, 188, **189**, 190
    Darwin's finches, 169–173
  sexual, 192
  species, 29
Semionotus, semionotids (neopterygian fishes), 80, *81*, 157
sexual selection, *see* selection, sexual
shadscale, *see Atriplex confertifolia*
shrew
  masked, *see Sorex cinereus*
  smoky, *see Sorex fumeus*
Siberia, 137
Silurian, **4**, 82, *87*, **88**
Simpson, George, 20–22, 25
snakes
  longevity, **175**
Solar System, 44–47, 89
solstice, *49*, 56
Sophora toromiro, tree (Leguminosae, Easter Island), 182
  pollen, *182*
Sorex, shrews (Soricidae)
Sorex cinereus, masked shrew (Soricidae, North America), *143*

Sorex fumeus, smoky shrew (Soricidae, North America), *143*
South America, 62
  glaciations, **88**
  vegetation change
    Quaternary, 124–126
speciation, 15, <u>16</u>, <u>18</u>, 19, <u>19</u>, 22, 23, **23**, 26, 27, <u>27</u>, 28, *28*, <u>28</u>, 31, <u>31</u>, *32*, 33, 34, 43, 152, 156, 157, 159, 175, 176, 189, 195
  allopatric, 18, 19, 24, <u>27</u>, 154, 188, 191, 193
  climatic change, 194
  corals, 173
  distribution change, 176
  environmental change, 176
  frequency, 191
  geographic, *17*, 18, 19, 27, 164, 165, **187**
  peripatric, 27, 32
  plant, 22
  rate, 18, 173
  sympatric, 18, 19, 24, 191
species
  as individuals, *see* individuals, 191
  as interactors, 152
  biological, 18, 24
  constant, 31
  duration, 40, 157, 174, *174*, **175**
  emergent characters, 176, 183
  evolutionary role, 19
  nature of, 175, 188
  origin, 8, 9, 14, 153
    mammals, *180*
  rates, 178
  sorting, 176, 183
  spatiotemporally-bounded, 31, 191, 194
  stability of, 8
species selection, *see* selection, species
spectral analysis, 67, *68, 69*, 72, 73, *73*, 77, 78, 84, *86*
Spermophilus tridecemlineatus, thirteen-lined ground squirrel (Sciuridae, North America), *143*
Spondylus gaederopus, bivalve molluscs, 148
squirrel
  ground, thirteen-lined, *see Spermophilus tridecemlineatus*
stadial, 94
stasis, 26, 154, 159, 165, 167, 175, 184, 188, 189
steppe vegetation, 96, 99, 100

240 · Index

Stipagrostis, desert grasses (Gramineae), 128
stratigraphic units
    usage, 3
struggle for existence, 21, <u>25</u>, **187**
Styloceras, Andean trees (Buxaceae)
    pollen, *125*
Sun, 44
Symphoricarpos, deciduous shrubs (Caprifoliaceae), 117
Symphoricarpos cf. *longiflorus*
    macrofossils, 118
Symplocos, tropical trees (Symplocaceae)
    pollen, *125*
Synaptomys borealis, northern bog lemming (Arvicolidae), *143*
Synorichthys, redfieldiids (chondrostean fishes), *80, 81*
Syracosphaera pulcha, coccolithophores, *146*
Systematics and the Origin of Species, 18
systematics, animal, 15
systematics, plant, 15
Syzygium, tropical trees and shrubs (Myrtaceae), 127
    pollen, *128*
Sørenson's index of similarity, 119, *121*

Tachinus apterus, beetles (Staphylinidae), 159
Tamias striatus, eastern chipmunk (Sciuridae, North America), *143*
Tanytrachelus, small reptile (protorosaur), *80*
Taxaceae, coniferous trees (gymnosperms)
Taxodiaceae, coniferous trees (gymnosperms)
TCT, coniferous trees (Taxodiaceae, Cupressaceae, Taxaceae), 110, 111
    pollen, *110*
tectonic uplift, 87–89
Tellina, bivalve molluscs
Tellina compressa, 148
Tellina pulchella, 148
Tempo and Mode in Evolution, 20
Tenaghi Philippon, Greece, *98, 99*, 96–102
Tertiary, **4**, 10, 61, 72–74, **88**, 179
Tethyan Ocean, 76
Thalassiosira graviola, marine diatoms, 147
Thelypteris, ferns (Thelypteridaceae)
    spores, *127*
Thuja plicata, western red cedar (Cupressaceae, North America), **151**
Thysanophora hypolepta, pulmonate molluscs, 140, 162

Tibet, 87, 137
Tilia, lime and basswood trees (Tiliaceae), 103, 105, **152**
    pollen, *105, 109*
Tilia cordata, small-leaved lime (Tiliaceae, Europe), **151**
time-scale
    geological, **4**, 41, 71
    radiocarbon, 42
    radiometric, 79, 82
time-scales
    ecological, 2, 38, 168–173, 188–190
    geological, 1, 2, 24, 27, 38, 152, 190, 191, 195
    Holocene, 42
    human, 1
    Milankovitch, 2, 4, 66, 72, 75, 85, 90, 92, 98, 102, 126, 154, 158, 173, 175, 176, 184, 188, 189, 191, 194–197
    Quaternary, 1, 41–43, 188
time-series, 65, 98, *108*
    geological, 67, *68*
toads, *see Bufo*
transmutation, 7
trees
    pollen, *see also* individual taxa, 97, *123, 124, 130*
Trema, tropical pioneer trees (Ulmaceae), **151**
trends, evolutionary, 27, *28*, 29, 31, <u>33</u>, 191
Triassic, **4**, 78, 79, *79–81*, **88**
Tribulus, usually shrubs (Zygophyllaceae)
    pollen, *130*
trilobites
    longevity, **175**
Triumfetta semitriloba, tree (Tiliaceae, Easter Island), 182
    pollen, *182*
Tsuga, hemlock trees (Pinaceae), 100, 111, 114, 181
    pollen, *99, 110*
Tsuga canadensis, eastern hemlock (Pinaceae, North America), **151**
    pollen, *113*
Tsuga mertensiana, mountain hemlock (Pinaceae, North America), **151**
Turkana Basin, Kenya, 163–164
turnover-pulse, 31–33, 194–195
Turseodus, palaeoniscoids (chondrostean fishes), *80, 81*

turtles, see *Clemmys*
TWINSPAN, *106*
Typha, perennial marshland herbs (Typhaceae)
  pollen, *130*

Ukraine, 159
Ulmaceae, tropical and temperate trees and shrubs (dicot angiosperms), 100
  pollen, *99*
Ulmus, elm trees (Ulmaceae), see also *Ulmus/Zelkova*, 96, 137, **151, 152**, *169*
  pollen, *97, 99, 109, 113*
Ulmus/Zelkova, see also *Ulmus* and *Zelkova*
  pollen, *75, 101*
ultra-Darwinians, 26
Umbelliferae, herbs (dicot angiosperms)
  pollen, *128*
Umbellosphaera irregularis, coccolithophores, *146*
uniformitarianism, 26, 34, 35, 38, 154
Ursus maritimus, polar bear (Ursidae), 157
Urticaceae, herbs, shrubs, and trees (dicot angiosperms)
Urticaceae/Moraceae
  pollen, *123*
Uvigerina peregrina, benthic foraminifera, 147

Valle di Castiglione, Italy, 96, *97*, 101, 102, **152**
Vallea, trees (Elaeocarpaceae)
Vallea-type
  pollen, *125*
Valvata, prosobranch molluscs, *163*
Variation and Evolution in Plants, 22
Vauquelinia californica, desert plants (Rosaceae, North America), 117
vegetation
  Miocene, 74
vegetation change, 127
and climatic change, 42
anthropogenic, 132, 136, 182
Holocene
  British Isles, 103–107
  Europe, 102–103, *106*
  North America, 111–115
  Sahara, *130*, 127–131
late-Quaternary
  Easter Island, *182*
  Japan, 137
  New Zealand, *135*
  North America, *119*, *121*, *122*
Quaternary
  Africa, 126–127, *128*
  Asia, 136–137
  Australasia, 131–136
  Australia, *131*, *134*
  Central America, 122–124
  Europe, 101–102
  Greece, 96–101
  Italy, 96
  New Guinea, *136*
  North America, 108–121
  South America, *125*, 124–126, *127*
vegetation, past
  modern analogues, 103, *108*, 115, 122, 132
Venus, 45
vertebrates
  distribution change, 141–145
Vrba, Elisabeth, 31

Wallace, Alfred, 6, 10–11
Weinmannia, trees (Cunoniaceae)
  pollen, *125*
woodrat
  eastern, see *Neotoma floridana*

Zelkova, shrubs and trees (Ulmaceae), see also *Ulmus/Zelkova*, 96, 100, 102, **152**
  pollen, *97*

DATE DUE

Demco, Inc. 38-293